The Digital Pandemic

The Digital Pandemic

Reestablishing Face-to-Face Contact in the Electronic Age

By Mack R. Hicks, Ph.D.

New Horizon Press
Far Hills, NJ

New Horizon Press
P.O. Box 669
Far Hills, NJ 07931

Hicks, Mack R.
The Digital Pandemic: Reestablishing Face-to-Face Contact in the Electronic Age
Cover design: Robert Aulicino
Interior design: Susan M. Sanderson

Library of Congress Control Number: 2009927401

ISBN 13: 978-0-88282-315-7
New Horizon Press

Manufactured in the U.S.A.

2014 2013 2012 2011 2010 / 5 4 3 2 1

To my wife, Susan, for her patience and support.

To Charlie Chaplin, Paulette Goddard and the Hollywood cast of *Modern Times.* We get it, folks…we get it.

Author's Note

This book is based on the experiences of the author and reflects his perception of the past, present and future. The personalities, events, actions and conversations portrayed within this text have been taken from memories, interviews, research studies, letters, personal papers and media-related articles and features. *The Digital Pandemic* is based on extensive personal interviews, research and insights from the author. Certain names have been changed and recognizable characteristics disguised except for those of contributing experts.

"Electronic technology is within the gates, and
we are dumb, deaf, blind, and mute about its encounter
with the Gutenburg technology on and through which the
American way of life was formed."– Marshall McLuhan,
The Gutenburg Galaxy: The Making of Typographic Man

Table of Contents

Introduction

We are in the midst of a technological revolution that affects every facet of our lives. Changes are coming at computer speed so fast, in fact, that some question if the information technology (IT) society is a blessing or a curse. By the time research studies are completed, the landscape has already been changed.

Despite protests from some nervous parents, cynical teachers and the general public, some experts tell us to relax. In an article by former *Wall Street Journal* publisher L. Gordon Crovitz, futurist Ray Kurzwell explains, "...the power of computers is just the latest example of more than a century of exponential growth in communication and information technologies, beginning with electromechanical power and continuing through vacuum tubes and transistors."[1]

Those who are middle-aged and older don't understand, many computer experts say. They call it a generational gap. According to such authorities, this generation better get with the program—and be quick about it. Multipurpose cell phones, computer games and Web sites such as Facebook and Twitter are teaching kids more and better than traditional schools, stimulating both creativity and social skills. Or so the computer proponents say.

Some people believe automation and the machine culture free us from mundane concerns over food, shelter, safety and the like.

Those who tout electronic devices theorize that now we can enjoy our "self-actualizing" potential for the first time in history. Concerns over machines will lessen as we take the high road to a world of creativity, empathy and self-awareness.[2] We'll control machines and machine-like thinking—not the other way around. You bet. But experience and scientific research doesn't always support these theories.

Not only do some psychological studies raise concerns over the "machine's" negative impact on individual learning and development, but the growth in government regulations, legalistic thinking, objective testing in schools, increasing dependence on electronic gaming and the slicing and dicing of humans into abstract categories all point to the power of machines to influence our minds and ravage our souls.

The machine culture leads to complex changes, but I contend that a disease is spreading—lock up the kids. Several books supporting the information technology explosion rely on broad-based surveys to make inferences about *cultural* changes. *The Digital Pandemic*, on the other hand, relies on clinical experience and psychological research to sort out the impact of technology on the *individual*. In this book we'll examine individual, social and, lastly, cultural possibilities. We'll discuss the impact of IT on people who face problems in the real world—not virtual ones.

We've all come across anecdotal reports and bits and pieces about information technology, and though we all have our own opinions, this book is an attempt to bring the disparate parts together and connect the dots. We want to get away from machine thinking that breaks subjects into easily digestible and unrelated parts and instead take a measured and comprehensive view of IT. Only when we confront its influence on many aspects of our lives *at one time* do we experience the full impact of its power and potential threat.

The IT movement affects each individual in different ways. This is due to gender, age and our unique personalities. We will look at two significant personality types to help us understand the growing impact of the IT movement.

Dividing individuals into two hypothetical groups—Gatherers and Hunters—we'll focus on these two transcendent personality types and show how they differ as well as their ultimate interdependence. Gatherers react and organize; Hunters act and create. Because machines affect people differently, it's important to learn how these two personalities influence our lives. Then we can begin to deal with the turbulent and rampaging machine culture. The differences between Gatherers and Hunters yield a powerful metaphor for understanding and interpreting the IT movement.

Uncle Gaderian, the pandemic robot, is a gatherer of kids. He stands at our front doors, offering gifts that dazzle and mesmerize. Our children will never be bored, he assures us. But will they still dream of lollipops and angels' wings? And where is he taking them, this Pied Piper in robotic armor?

Illustration courtesy of Rebecca Skelles

"Welcome to My World"

1

How We Got Addicted

Influences of mechanization.

The encroachment of mechanization is destroying the human spirit. We have seen the relentless movement inward from Hunter and Gatherer to agricultural farmer to industrial worker to information economy to office automaton to post-modern thinker to cut and paste modern browser. The question we need to ask ourselves is: Are we moving from active, sensitive and creative to passive, mechanical and conventional?

Are we destined, like futuristic space aliens, to have massive heads, Lilliputian bodies and squeaky voices? Or is this a man-made problem? We will either use technology or it will use us.

We'll examine the hypothesis that machines are destroying our souls. It's not just the computer, the Internet or electronic games, but the mechanical, detached and sedentary approaches to life that are troubling…and the thinking and behavior they encourage at the expense of emotion, creativity and interpersonal awareness.

Where are these influences evident? Smart phones, televisions, computers, electronic games, YouTube, MySpace, Facebook, objective testing, airport body scans and categorical thinking that reduces us to mere abstractions (and that's just the beginning). What about billboards at malls that spy on people and change the product signage to fit the shopper's profile? The dry, analytic model of science is expanding its influence into discussions about final truths and trying to answer the "why's" and not just the "what's." Mechanical medical devices, such as computer-driven MRIs (magnetic resonance imaging) and electronic palm prints, are used for identification, research and our good health. But where does our spirit go after we're nicely sliced and diced?

Are we unique and unpredictable humans or are we mere abstract concepts, like middle class, bipolar or urbanite? Science lacks a value system, unless you consider maximizing the survival of the species and the unfettered investigation of everything *as a value system*. No one's at the helm of the science ship and most scientists like it that way. How can they discover new directions if they follow old paths? Some people in the digital generation seem to agree. We don't need captains, helms or road maps, according to the electronic gurus. We can play it by ear and be the better for it: post-modernism with a hard drive?

According to these experts, technology and science aren't perfect. They say such fields have a pretty good track record of helping our lifeblood, except they also have created weapons of mass destruction. Maybe when science reaches its ultimate goals, we won't need such old-fashioned terms as emotion, conscience or free will. Science will redefine us and in the process do away with the old mythologies. If we have to compromise here and there, it's worth it—so they say.

In this book I differ with such a mechanistic view of human beings and advocate retaining the advantages of technology, but not, however, giving up the basics that make life worth living. Things like valuing people regardless of status, background or helplessness; enjoying the simple pleasures of family and work; walking on green spongy grass; playing catch with a real ball and having to chase an errant throw into the weeds; meeting people face-to-face; learning through writing and composition; nurturing, touching, feeling and experiencing the pressure and smell when you press an old-fashioned crayon onto *and into* paper; and human dignity, philosophy and faith in things that can't be measured.

Moreover I believe that mechanization is influencing our thinking and communications in many ways that are not positives.

What will occur when our present six-year-olds, whose brains are being shaped to conform to the IT model, hit the workforce, the parent-force and the citizen-force? What then?

High-tech generation.

Our young high-tech generation has shorter attention spans, especially when they enter the regular school environment. They find television boring, so about a third of them use other media (especially the Internet) *while* watching television. Young people today spend much less time reading for leisure than ever before. Why use your own imagination and move *slowly* through an intricate plot when you can be blasted with stunning audio and visual effects that rivet you to "the game"?

Recent research by Professor of Clinical Psychiatry Gary Small, M.D., and his team indicates that technology is changing our lives but also changing our brains.[1] The human brain is quite malleable and constantly changes in response to the environment. A young person's brain, which is still developing, is particularly vulnerable. It is also the most exposed to new technology. Information technology (IT) natives—young people born into the world of laptops, cell phones and text messaging—spend an average of eight and a half hours each day exposed to digital technology. These exposures rewire their brains and neural circuitry. All that tech time may diminish "people skills, including important emotional attitudes like empathy."[2]

Today, parents receive some guidance to help determine the desirability of electronic game *content*, but there is no guidance when it comes to the *process* itself. The content of a basic video game may be entirely innocent, but the *process* of spending many hours playing that game may create learning and attention problems that won't show up until later.

Social animals.

In a *Wall Street Journal* article, Harvard psychology professor Daniel Gilbert says, "Human beings are social animals, so it is no surprise that our greatest sources of happiness and unhappiness are our social relationships. When it comes to happiness, consumer goods that involve our friends are better than those that replace them."[3] New studies show the negative effects of online "gaming" on self-awareness, understanding other people and comprehending our surrounding environment. This leads to decreases in adolescent life satisfaction. [4]

Young people across the country are "...posing like statues in public squares, dropping their pants in train stations and bursting into song in malls....15 pairs of identical twins, dressed in identical outfits, filled a New York City subway car and mirrored each other's actions, without explanation," writes journalist Ellen Gamerman in *The Wall Street Journal*.[5] Why are these "urban alchemists" so desperate to create magic in their lives? Is it political protest and rebellion against authority, reminiscent of the 1960s, or is it just people who want to *feel something* again?

Are tattoos and flesh jewelry a way of saying, *Hey, look at me, I'm important. I'm a human being, not just a mechanical part*? And when pants drag a little, is this rebellion or just teen style? Maybe teen bad manners are partially caused by being machine-raised. Maybe our friendly and sympathetic machines will pacify our misplaced desire for fresh air. Perhaps video game console makers want us to do more than just play online—they want us to live in virtual worlds.

Virtual reality.

One electronic company promises a 2009 online world for console owners: "...a virtual-reality environment where gamers with custom-made avatars can browse a virtual shopping mall, shoot pool with friends or strike up a conversation at an online town square before linking up to play games together online"[6] as reported in *The Wall Street Journal* article titled "A Way for Gamers to Get a Life." (An avatar is a computer user's representation of himself or herself or of an alter ego in various forms.) While there is no charge for "...standard clothing and a furnished apartment or personal space," the company wants to "...sell specialized clothing and unique furniture..."[7] for a small fee.

We live at a time when we can go to the shopping mall, buy furniture for our second home or party with friends without ever going outside. As documented in an article by Alex Roth and Paulo

Prada, excitement is brewing in Georgia about a "...15,000-square-foot-building [that] will feature interactive exhibits, including fishing simulators where visitors in a fake boat can struggle against computer-controlled fish. Outside, a wooden path will lead visitors through simulated Georgia topography—from mountains to Piedmont to swampland—as trout, bass and bream swim in aquariums and pools along the way."[8]

The awful truth.

What about the addiction process and its effect on self-concept, emotional growth and happiness? An anti-tobacco commercial features a smoker who struggles to speak through a loud, persistent cough. He refuses to be "a quitter" and give up his self-destructive habit. Tech addicts can rationalize with the best of the tobacco slugs. "Look, man, I spend my days working in an isolated cubicle with no chance to socialize. (He's saying he's a machine part, but he doesn't know that.)

"On the Internet I can meet friends and watching television relieves a hell of a lot of stress. At least I'm not on drugs and I don't gamble my money away. Give me a break, man, and let's move on into the new world." Ironic, isn't it? Machine thinking and technology are causing this poor guy's stress, and now he's turning to machines for the cure.

Of course, he is giving up more friends, greater success, increased income and better physical and mental health. Too bad he doesn't know the awful truth. He can still enjoy his machines as long as he doesn't *become* his machines.

In this book I will show electronic addicts and their families how to break hardware addiction and tame thirst for "computer time." I will also try to reignite latent curiosity about people and show how to use natural skills of observation to understand ourselves and others.

A powerful tool.

Research is a powerful tool for reform. It seems logical that the Internet and computer games would increase attention and the ability to focus, but much recent research disagrees. Tasks requiring high levels of direct attention can cause the attention system to fatigue. As reported in the *American Journal of Public Health,* Frances E. Kuo, Ph.D., and Andrea Faber Taylor, Ph.D., conducted a study showing that increasing children's time in natural green settings diminishes symptoms of inattention.

Participants in the study were 452 parents of children ages five to eighteen who had been formally diagnosed with attention deficit hyperactivity disorder (ADHD) by a physician, psychologist or psychiatrist. The parents were diverse in terms of their socio-economic statuses, the ages of their children, the community types in which they lived and the regions of the country from where they were came. The severity of their children's ADHD symptoms was also diverse, and both males (80 percent) and females (20 percent) were represented.[9]

"Parents nationwide rated the aftereffects of 49 common after-school and weekend activities..." in terms of whether it made their children's symptoms much worse than usual, worse than usual, same as usual, better than usual or much better than usual, for an hour or so after the activity ended. Activities were described as occurring in either green outdoor settings such as parks, farms, green backyards, neighborhood spaces, built outdoor settings like parking lots, downtown areas or indoor settings. The parents knew nothing about the purpose of the study. The parents' ratings indicated that the children's ADHD symptoms were significantly better after participating in green outdoor activities compared to activities that occurred indoors or in constructed outdoor spaces.[10]

The study also controlled to ensure that this outdoor fun wasn't the result of more activities taking place in a green space. Results from a supplemental analysis indicated that even when the activities were identical, there was a greater reduction in attention problems when they took place in a green space.

Abraham Lincoln wasn't the only United States president who grew up in the outdoors. In his book *An Hour Before Daylight*, Jimmy Carter says, "My most persistent impression as a farm boy was of the earth. There was closeness, almost an immersion, in the sand, loam, and red clay that seemed natural, and constant. The soil caressed my bare feet, and the dust was always boiling up from the dirt road that passed fifty feet from our front door."[11]

Reading? Who needs it?

Does the IT explosion leave any room for reading? In 2004, The National Endowment for the Arts in Washington, D.C., reported that fifteen- to twenty-four-year-olds spent an average of seven minutes reading on weekdays; people between thirty-five and forty-four spent twelve minutes; people sixty-five and older spent close to an hour.[12]

In an article from *The Wall Street Journal*, Amazon CEO Jeff Bezos states that, "Laptops, Blackberries and mobile phones have shifted us more toward information snacking, and I would argue toward shorter attention spans."[13]

Internal and external protests.

When we spend too much time on machines, even our bodies protest. Sometimes they stiffen up and stop working. There's something about our physical reactions to computers that are definitely negative. According to Melinda Beck, in her article "When Your Laptop is a Big Pain in the Neck," long bouts on the computer and incorrect positioning can result in "...headaches, pains in the temporo mandibular joint (TMJ) and carpal tunnel syndrome, in which pressure on wrist nerves causes tingling and numbness in the hands," in addition to "...pain and stiffness in the neck, shoulders, back and arms."[14] (Doesn't do much for your posture, either).

And sometimes we protest intellectually, as well. How many curse words have you leveled at your machine? Has any human boiled your blood or elevated your blood pressure like your machines? Wait until it is fully running your life, parceling out candy and other rewards, along with positive words in that flat, monotonous voice just—to—keep—you—going. The next time you walk into your office or open your laptop, glance out of the corner of your eye. You might catch your machine staring at you. And someday it will. But when it finally decides it doesn't need you, you'll hear the click of the delete button.

Electronic relationships.

Hal and Gail are tooling down the highway from their son's summer camp in the family sedan when Hal complains to his wife that her onboard GPS is too loud. "Besides," he says, "my cell phone has GPS with turn-by-turn navigation and it's only four inches long."

"Don't get me started," Gail shoots back. "My new cell phone has twice the speed at half the price. I know yours has superior integration and a secure e-mail platform, but you can keep the applications." Quiet in the backseat causes Gail to turn and look at their two children, who sit side by side typing away.

"Who are you guys e-mailing?" Gail asks.

"Only old people e-mail," her son snaps. "We're texting a bunch of people; mostly each other," he says, in a nearly complete sentence.

Gail shakes her head. "So, to summarize, you two haven't seen each other in two weeks and you're sitting one foot apart—texting?" Silence greets her again. She shakes her head and turns to her husband for support, but he's listening to music on his stereophonic earphones and isn't in a response mode.

Online vs. face-to-face.

Relationships are based on trust, empathy and imagination, none of which are strong points for a computer. Computers can collect data and produce incredibly useful information in a millisecond, but they can't recognize basic facial expressions such as anger and happiness. We seem to be getting "relationship Lite," not the real thing.

Founding executive editor of *Wired* magazine, Kevin Kelly, in his book *Out of Control*, predicts living organisms will become the model for man-made systems. "Control is out, out of control is in."[15] In *Playing the Future: How Kids' Culture Can Teach Us to Thrive in an Age of Chaos,* author Douglas Rushkoff tells us to embrace the new technological age and challenges us to accept kids as the latest model of human beings, "equipped with a whole lot of new features."[16] He believes kids relish a technological climate of chaos and go with the evolutionary flow, while their elders cling to obsolete institutions.

In *Born Digital: Understanding the First Generation of Digital Natives,* law professors John Palfrey and Urs Gasser state that many aspects of the way "Digital Natives" lead their lives are cause for concern, but "…parents and teachers need to let Digital Natives be their guides to this new connected way of living."[17] Business executive Don Tapscott's book, *Grown Up Digital: How the Net Generation is Changing Your World,* concludes that "Net Geners" are smarter and quicker than their predecessors. Rather than meeting face-to-face, Tapscott's team interviewed nearly 8,000 young people through *online questionnaires.* Tapscott's study is endorsed by twenty-one business leaders, including the CEO of Internet search engine Google and the editor-in-chief of *Wired.* Fifty companies collaborated in the study. "The reports are *proprietary* to research sponsors, but *some* of the high-level findings and main conclusions can now be shared."[18] (My italics)

In my opinion, many of the studies share a common flaw: their descriptions of people are devoid of such considerations as genetics, temperament, gender, personality style or individual differences. There were some people who believed the automobile would never replace the horse and buggy. It did, but it's too bad it took almost one hundred years to require the use of seat belts. Talking on digital telephones while driving a vehicle causes more accidents and results in more deaths than drunk driving, but so far the phones are still conveying information critical to good driving, such as, "Do you want asparagus or green beans with the lamb stew?"

Anthropologist Mizuko Ito at the University of California, Irvine, led a group of anthropologists on a fifty million dollar *ethnographic* study for the Pew Internet & American Life Project. They looked at social and recreational aspects of technology from a youth perspective. The study revealed some important information on what kids said they were doing on a day-to-day basis within our culture. Ito's study was an important foundational work, but it relied on young people's own reports and opinions rather than outcome studies. It did not focus on digital usage in our schools, and parent and teacher perspectives were not assessed. [19]

In the twelve to seventeen age range, 97 percent said they played video games and 82 percent said they played alone, occasionally. What about mature and adult games? Twelve to fourteen-year-olds are equally inclined to play mature and adult games as older children and young adults. Ito used students in the range of eight to twenty years old, which means that some of the students were ten or twelve years of age before much of the current technology was introduced. These older students' brains were not exposed to the early stimulation that is of concern in some neuropsychological research today. As mentioned before, it's difficult to research the new technologies because of their rapid changes. Also, the researchers couldn't study the effects of such widely used devices, such as video game consoles, cell phones and the like. According to the researchers, lower-income families couldn't afford them.[20]

Think IT learning doesn't create strong attachments? Ito and her colleagues interviewed hundreds of students across the United States. One twenty-one-year-old had discussed his all-consuming relationship with an electronic game involving a future cataclysmic civilization. He said it was very mechanistic and repetitious, with an emphasis on rational thinking. He described it as "meticulous problem solving" with "frighteningly gorgeous environments."[21]

Privacy, trust and control.

Is the idea that machines may be gradually taking us over and infecting our minds and hearts real paranoia? Could be. What about the new security machines that strip us bare at airports? The bodies of men, women and children are exposed while disinterested security agents look on. Urban myth has it that Homeland Security screeners like to select attractive women for body scans. Are we pushing the envelope when it comes to human dignity? Could that old-fashioned dignity idea become a thing of the past?

What about privacy? That lady with the cell phone can take your picture or transmit your actions to her friends in other parts of the world. Recently, in a *St. Petersburg Times* article, an Olympic gold medalist was seen smoking marijuana from a bong/pipe on a cell phone camera picture.[22] Newspapers later printed the story worldwide. There's been some debate about the hoards of data shared by 200 million users of Facebook. In an article by Jessica E. Vascellaro, published in *The Wall Street Journal*, she writes that "...consumer-advocacy blog Consumerist.com posted an item... with the provocative headline 'Facebook's New Terms of Service: We Can Do Anything We Want With Your Content. Forever.'"[23]

High tech helper or big brother? Todd Lewin reports for the *Associated Press* that two employees of a surveillance equipment company had glass-encapsulated microchips with miniature antennas embedded into their forearms.

"We're really on the verge of creating a surveillance society in America, where every movement and every action—some would even claim, our own very thoughts—will be tracked, monitored, recorded and correlated," says Barry Steinhardt, ACLU, Washington, D.C.[24]

Trust: Where did that go? Dressed in full armor, a knight of old opened his visor to show others his face; this way people could identify him as friend or enemy. The raising of the visor, usually with the right hand, evolved into the military salute, which carried the same significance. This was replaced by the handshake, which showed that the fist didn't conceal a weapon.

Today's biometric technology is replacing these civilized gestures by reading the veins in our hands. You simply place your palm on the scanner and the Great Oz will know your age, employment—or lack thereof—government identification number (Social Security), spouse, family, reliable acquaintances and physical conditions.

In an article written for *Harper's Magazine*, a pilot flying a B-2 bomber from Missouri to Baghdad and back states, "Technology trumps our shared human nature. I tell myself that my actions will help save the lives of soldiers who are racing north out of Kuwait. This is honorable. It is not honor." This intelligent and sensitive pilot asks: "Will I kill to free another man's slave into a world that may be more chaotic, anarchic and dangerous?" and observes "It is difficult to match deeds with the ancients. I am cloaked in the *conceit of technology*."[25] (My italics)

We used to think pilots were protected from the brutality of war because they didn't engage in trench warfare and man-to-man combat. Now, technology has induced significant stress on air national guardsmen who operate bomber drones over Iraq via remote control from the safety of Southern California. In an article published in the online magazine Slate.com, writer William Saletan says that some of these drones have equipment that allows guardsman to see through walls. The effect of these devices, in the words of one former U.S. military official interviewed by the *Los Angeles Times,* is that insurgents "are living with a red dot on their head."[26] Drone pilots sitting in the comfort of their air-conditioned offices watch 500-pound bombs all the way to impact *and,* unlike the view from a traditional plane, they can see the resulting fatalities in high-resolution detail.

Mechanical thinking.

Many people are concerned about the *contents* of computer games or what dreadful excuses for education or entertainment are showing on our television sets. Granted, content is important and can be damaging to individuals and our society. My concern is the insidious *process* of mechanical thinking.

Recently, I went to a hospital for some blood work and when I entered the laboratory waiting room, four people were seated behind an L-shaped counter, ready to greet me and take care of me (or so I thought). No one else was in line. Fronted by computers, their eyes held steady to their respective screens. They were dressed in casual enough garb. No one looked up; no one gave me a single sign of recognition, not even a fluttering eyelid. *What are they doing?* I wondered.

I noticed a yellow business pad for names and times of entry. Assuming it was the sign-up sheet, I signed in and sat on a chair only a few feet from the counter. Fifteen minutes later, I became impatient and approached the dynamic foursome again. No one else had inked the sign-up sheet. After getting the attention of the closest employee by coughing and finally waving my hand, I asked her whether I had followed the correct procedure. She blinked and stared at my chest (later we'll discuss *microinequities*—she was a master). Yes, she nodded. "Go to Station Two." She disappeared back inside her white-lighted screen and world of ordered chaos. I looked for Station Two.

About four feet away, on the vertical part of the L-shaped counter, was a paper with the number 2 scrawled on it. This lady didn't look up either, but when I dangled the physician's order for blood work over her computer, she took it and looked up. "Are you here for blood work?" she asked. I said yes. She pointed to a wall and told me to wait over in that direction. Before long, a sunny-faced medical technician with an impish smile bounced into the waiting room. The computer appendage handed him the physician's order, but said nothing. He had to walk among the patients, calling out my name.

Later, I discussed my experience with the personable medical technician. He reported that the people at the front desk usually gave him no information about the patients and more likely than not handed him the doctors' orders with the blank side up. He was obviously unhappy with the situation but pleased to tell me that the hospital was highly responsive to patient feedback. The med tech gave me an observation card and indicated that the woman in charge of public relations was sensitive to patient needs and would appreciate any comments I might make.

It wasn't until I got home that I discovered the card had already been filled in by another patient, giving her address, phone number and other information I didn't need to know and shouldn't have had. I filled in the evaluation card, making it clear that my evaluation was separate from the person who had filled it in originally. I never heard anything from the hospital. Two months later I received a call indicating that I owed money for my blood work, despite the fact that my secondary insurance covered the entire cost.

We question the value of machines for teaching, but surely they help us sort out records. Right? Some people question whether digitizing medical records adds too much complexity to the system. Dr. Scott Silverstein states in *The Wall Street Journal*, "For 12.7 billion pounds, the U.K., which already has socialized medicine, still does not have a working national HIT (Health

Information Technology) system, but instead has a major IT quagmire. HIT (with a few exceptions), is largely a disaster."[27] A *Wall Street Journal* article titled "Computer Spies Breach Fighter-Jet Project," reports that "...computer spies have broken into the Pentagon's $300-billion Joint Strike Fighter project, the Defense Department's costliest weapons program ever...computers used to control the U.S. electrical-distribution system...have also been infiltrated."[28]

But won't technology help reduce the worldwide problem of carbon emission? Peter Hopton, Chief Executive and founder of Very PC, would like to lower the carbon footprint of IT without sacrificing performance. Writing in the RSA Newsletter (Royal Society for the Encouragement of Arts, Manufactures and Commerce), he states, "One server in the data centre has a similar operational carbon footprint to a range rover sport [car] and IT worldwide is responsible for similar *annual emissions as commercial airlines 4 times over.*"[29] (My italics)

Love at first click.

One day I received a recorded message from an insurance company that gave inaccurate information. I called back to clarify the errors, only to reach a recording. This robotic voice kept asking me questions but seemed to have difficulty understanding me. I knew what the machine wanted, so I began to speak like a robot myself. I shaped my mouth in a stilted and unnatural manner to articulate. Once I did this, the robot rewarded me with clear directions, but did not seem one bit relieved or grateful. I was congratulating myself on my success when I realized that I was gradually being shaped into the automated IT world.

Another time I dialed 411 for a phone number. A seductive and lyrical, if somewhat cloying, voice filled me in on other available services. But she couldn't compete with the young lady who answered my call to a printer company for technical support. She was so personal, I swear she knew me. She warned me that she would be asking a few questions and admonished: "Be sure and speak up so I can hear you." *Yes, I'll try,* I thought. Later she made my spirits soar when she interrupted my mutterings with, "You're doing great."

Our cold-blooded friends don't really know us, but at least we can rely on their flawless logic. Right? I was having trouble printing a letter and got this message on my computer screen: "The printer has not yet responded, but the server may be able to proceed without the printer information. Do you want to continue to wait?"

Yes ☐ No ☐

I was able to get a real live person when I called my insurance company to warn them that their computer needed counseling. I was getting two to three reports, covering the same material, each week. I suggested a vacation for Uncle Gaderian, the pandemic robot—he seems to be rusting up and could use at least a lube job. The real person—I think—apologized and said she'd like to stop the mailings, but the computer wouldn't cooperate. She said, "I'm sorry, sir, we can't *suppress* it."

Here today, gone tomorrow.

Unfortunately, Digital Natives may not realize what they're missing in the way of personal service. It would be helpful if they came to recognize the losses as well as the gains associated with information technology. Passing information from generation to generation can be good. However, it took me years of training and thirty years in private practice to develop interviewing techniques that have proved *relatively* valid and reliable. I've found observing behavior is more accurate than asking people questions.

Young people are likely to be wary of strangers who ask questions about their increasing involvement with electronics and computers. This is part and parcel of our separate adolescent culture. Teenagers may hide negative aspects of technology and exaggerate positive experiences.

The new imaging machines also help in the dehumanization of the individual. Much brain research today starts with a person lying immobile, like a machine part, inside an MRI, which is a narrow metal capsule where scientists study what part of the computerized brain image lights up (in living color) when the subject solves problems or entertains certain thoughts.

Some scientists tell us that the whole is not greater than the sum of the parts and personality is merely an aggregate of carefully identified brain modules. How these modules interact in a cohesive way, and where they'll lead us, has not yet been sorted out.

Personality...personality.

In the next chapter we're going to examine the behavior of a couple of hypothetical personality types to better understand the impact of IT on people.

In the *Journal of Personality Assessment*, Dr. Theodore Millon writes, "No longer was personality to be seen as an integrated gestalt, a dynamic system comprising more than the mere sum of its parts. The pendulum swung toward empiricism and positivism; only 'observable' facts were in the ascendancy...this new breed of quasi-empiricist made a shambles of the inspired 'personality-as-a coherent-whole.'"[30]

This piecemeal approach to research is definitely working from the bottom up and hopefully will contribute to coherent systems in the future. In the meantime, to better understand the impact of information technology, let's plunge ahead and divide individuals into two personality types: the Gatherer and the Hunter. According to the article "Hunter Gatherers" by K. Kris Hirst on About.com, anthropologists use these terms to "describe a specific kind of lifestyle, that of all human beings until the invention of agriculture about 8,000 years ago...Hunter-gatherers hunt game and collect plant foods (called foraging) rather than grow or tend crops."[31]

How about our so-called "Cyber Generation", which spends a lot of its time browsing the Internet, foraging in the wilds, refusing to plant, cultivate and harvest crops? No, they don't need to struggle with long-range plans, mentors or proven methods of food production. They'd rather forage on their own, relying entirely on their own abilities. *Foraging*, as defined by Merriam-Webster's Collegiate Dictionary, Tenth Edition, is: 1. "food for animals, especially when taken by *browsing* or grazing." 2. "wander in search of forage or food."[32] (My italics)

In borrowing these terms from our anthropology friends, I'm defining the Gatherer as the realistic and conventional one, who is a linear thinker and relies on logic and objective information. The Hunter, on the other hand, is more intuitive, creative and more physically agile. In the next chapter we'll home in on their characteristics and encourage you to discover these personalities in yourself, your friends and your co-workers. We will study the effects of IT on Gatherers and Hunters and observe how interesting real people are. We'll look at a number of areas where the mechanization culture shapes our lives. Are machines the problem or are they harbingers of things to come? If we don't open a path that allows for both mechanization and humanization, we could easily "drift" away, into the not so starry night.

2

Halvin and Garrison

Hunter and Gatherer Personalities

"We have met the enemy and he is us."—*Pogo*

Let's look at how mechanization affects us individually, on a day-to-day basis. *We need to remember that our body is a machine*, but it's a superb machine. That's because of the attributes of speech, self-awareness and emotion. If mechanization is affecting us in a negative way, we first need to ask *who* we are. Since each person is unique, technology will affect every person in dramatically different ways.

Our goals of this chapter are twofold: generate interest and excitement in direct observation of real people and their behaviors (not machine-driven, i.e. *Survivor* and other "reality" television shows) and familiarize the reader with the Gatherer and Hunter personality types. Most people have inklings that their friends and associates have differing personalities, but sometimes it's confusing and overwhelming. By fitting everyone into two general types, we can learn to pick people out by their behaviors. Then we'll be in a position to better measure the potential impact of information technology on individuals.

I'm using the labels Gatherer and Hunter, because I have seen these two personality types in my private practice on a regular basis. Many of the characteristics described are found in personality testing and neurological research.

For clarification purposes, we'll divide our IT people into these two broad camps. You may find your place in either or both camps. As seen in the previous chapter, anthropologists use these terms to describe a specific lifestyle, that of humans up until the invention of agriculture around 8,000 years ago. Hunter-gatherers hunted game and collected plant foods by foraging rather than growing or tending crops. These typologies are found in modern theory, personality testing and brain research.

Descriptive labels for Gatherers include Judging and Thinking,[1] Realistic-Conventional,[2] Conformance and Self-Control,[3] Sequential and Linear, Logical, Objective, Focused, Categorical and Reliable. Gatherers would be the ego in Sigmund Freud's formulation and the adult in Transactional Analysis. Neuropsychological research would point these individuals in the direction of left hemisphere (or left-brained people).

Descriptors for Hunters include Sensing and Intuitive,[4] Creative Potential and Artistic,[5] Social Presence,[6] Simultaneous, Holistic, Spatial, Impulsive and right brain research. In Freud's formulations, the Hunter might fit into the id and superego categories while in Transactional Analysis the Hunter would fit the parent and child. Neuropsychological research would reveal characteristics similar to the so-called right-hemisphere person.

GATHERER	**HUNTER**
LITERAL	IMAGINATIVE
CONTENT	PROCESS
SEQUENTIAL	SIMULTANEOUS
REDUCTIONISTIC	EXPANSIVE
LINEAR	NON-LINEAR
BOTTOM-UP THINKING	TOP-DOWN THINKING
SEQUENTIAL PROGRESSION	PRIORITIZE/PROPORTIONALITY
LOGICAL PROGRESSION	EPIPHANY/SHIFT/INTUITIVE
LANGUAGE	ACTIVITY/MOVEMENT
FACTS	STORIES AND METAPHORS
PATIENT	EASILY BORED
CONTROL	SERENDIPITY

Mechanization poses the greatest immediate threat to the Gatherer because technology flatters the Gatherer's logical and conventional style. The Hunter's active and intuitive style is also attracted to electronic games such as first-person shooters and other adventurous formats. And, as we'll see, the Hunter can benefit from the organizational aspects of mechanization.

Powers of observation.

A second goal of this chapter is to tune up powers of observation following a Hunter and a Gatherer at home and in the office. Machines are full of bluster, promising immediate gratification, but what people really want is right under their noses—and it's free. Most people enjoy televisions, computers, the Internet and other high-tech applications, but sometimes they want more. That's where the real world comes in. Watching humans in 3-D and in real time can be very instructive. But remember, maybe they're watching you, too.

Let's look at speech, voice, gait, facial expressions and patterns of behavior to improve day-to-day observation skills and show how a computer personality (Gatherer) differs from computer hardware. Observation unmasks minds without reference to psychology books, the Internet or the online encyclopedia Wikipedia. Noticing bits of behavior that form patterns and, ultimately, personality styles will bring insight based on these observations and help to better explain personalities and motivations at home, in the office and in the community.

Do IT techies really look at people or do their eyes follow the perimeter of the visual field as reinforced by playing electronic games and exploring Web sites? And if they do slow down long enough to look at individuals, do they even have the vocabulary to describe what they are seeing? How many of you regularly look for the facial features described by these words: chiseled, sharp-featured, high-faced, wizened, ruddy, inscrutable, pinched, craggy, spectral, bloated, impish, cherubic and saturnine.

An example of insightful description comes from Aldous Huxley's book *Chrome Yellow*. "Next to Mary a small gaunt man was sitting, rigid and erect in his chair. In appearance Mr. Scogan was like one of those extinct bird-lizards of the Tertiary. His nose was beaked, his dark eye had the shining quickness of a robin's. But there was nothing soft or gracious or feathery about him. The skin of his wrinkled brown face had a dry and scaly look; his hands were the hands of a crocodile. His movements were marked by the lizard's disconcertingly abrupt clockwork speed; his speech was thin, fluty, and dry."[7]

Another telling description is from John Fowles's *The Ebony Tower*: "He advanced, hand outstretched, in pale blue trousers and a dark blue shirt, an unexpected flash of Oxford and Cambridge, a red silk square. He was white-haired, though the eyebrows were still faintly gray; the bulbous nose, and the misleadingly fastidious mouth, the pouched gray-blue eyes in a hale face. He moved almost briskly, as if aware that he had been remiss in some way; smaller and trimmer than David had visualized from the photographs."[8]

Like trash compacter WALL-E, from the blockbuster movie by the same name, we need to fight back and use all five senses to rediscover important relationships. Multitasking and machinery causes our senses to be immobile and allow others to see for us while promising better living through the Internet, but if we can learn the process of observation, we'll discover that what we really want is maybe right in front of us.

Odd couple.

When we look at the extreme levels of the Gatherer and Hunter personalities, we realize just how different they are. They think, listen and speak differently; they don't get along very well. These levels seem to ignore each other, because their differences make communication difficult. The Gatherer speaks and thinks in a detailed manner from the bottom up, while the Hunter thinks and speaks with a large brush, from the top down. Of course we probably all have aspects of these personalities within us to some degree and sometimes we can feel a tug-of-war within us.

Garrison, a thirteen-year-old Gatherer, and Halvin, a twelve-year-old Hunter, were in a video store checking out movies when they spotted a video that needed to be replaced. Garrison, the Gatherer, looked at the title and then gazed down the row of alphabetically arranged videos, checking the title against the alphabetical signage. Halvin, the Hunter, walked down the aisle looking for a space between the videos. They both arrived at the correct solution about the same time, but took much different routes to success. Consistent with their personality differences, Gatherers feel more comfortable with letters and words, while Hunters prefer spatial approaches.

Some ten-year-old boys with whom I talked had consistent differences between what I perceived to be Gatherer types and Hunter types. The Gatherers were fascinated with the electronic gadgets themselves. They rattled off every possible function built into the machine and even after an hour or two of non-electronic game activity, they came back again and again to repeat capabilities and report new applications. The boys seemed totally absorbed with the machine itself. Joe Morgenstern, film critic for *The Wall Street Journal*, reviewed the movie *Hotel For Dogs*. He wrote, "...as a mixed breed of sweet fantasy and rabid commerce, a film that grabs its audience like a chew toy and doesn't know when to let go."[9] These Gatherers don't just admire their machines—they can't let go.

The boys were enthralled with games where the format is to avoid barriers and let an object, such as a monkey or an automobile, move to its destination. It probably takes a mental age of four years or less. But the gamer has to *tilt* the floor up and down and from side to side to avoid the barriers. It's similar to tilt games from the old days that were constructed of wood and had holes in the floor. The object was to roll a marble past the holes to the goal. This exercise could be a "rush" for a Gatherer who may lack spatial-motor dexterity. In another game, where invaders tried to run through a maze to defeat the gamer, the player would select myriad defenses to attack the invader, then sit back and savor the neat graphics that proclaimed a mighty victory. Here again, these children didn't want to let go. Unlike kids in earlier times, these boys were not into sharing and the games seemed more important than the player or the interviewer.

The Hunters were different—they weren't unduly impressed with the technology. They focused on a more assertive approach and more complex and adventurous games. Tilt coordination wasn't as appealing and they were less "stuck" than the Gatherers. Though they did share one thing with the Gatherers: they were not into sharing.

The movie actor Dennis Quaid reported that his twin children almost died due to an accidental overdose of medication. He recommended that hospitals color code all medication containers to prevent these kinds of mistakes. This is definitely a solution for the Hunter, who will respond to color, but some Gatherer nurses will respond more accurately to written words than to color coding. So personality is important. It's easy to see why an actor, who responds so well to visualization, would assume that color coding would help everyone, and it may, to some degree.

The gals have it.

Why are the vast majority of female television news announcers gorgeous twenty-somethings? They must be smarter than the plain-Jane and bespectacled applicants from MIT and other universities. Some of these women must have interviewed well.

Ginger, a Montreal television announcer, reported on yet another loss for the storied Montreal Canadians hockey team. But she never compromised her upbeat countenance and radiant smile. The hockey story was a disappointing one, but her voice continued to project an automatic tone of frozen happiness, even while later reporting on a tragic trailer park murder. Her smooth as a river rock, implacable voice is consistent with the Gatherer personality. Television announcers, commentators and reporters often have a heavy dose of Gatherer personality because of their love of words, reading and writing.

Hallie, an interior decorator, told her friend Sally that a Gatherer accountant had misidentified Hallie, thinking she was Sally. But Hallie couldn't remember the name of the accountant. Hunters are often superior in visual perception but have difficulty with names, while Gatherers have a good memory for names but are weak in visual perception. Hallie, the Hunter, couldn't remember a name, and the accountant, a Gatherer, couldn't maintain a visual image of Hallie. Hallie's an artist and enjoys listening to Garrison Keillor's radio show, *The Prairie Home Companion*, but sometimes can't remember Keillor's name. She remembers the content and humor and has a pictorial concept of Lake Woebegone in her mind, but has difficulty with the letters that comprise Garrison Keillor's name.

When Ginger, a Gatherer, goes to an art museum, she enjoys viewing the paintings up close. She likes to see the brush marks and enjoys the pointillism technique where tiny dots are combined to form a picture. She also likes descriptions of the artist's biography, place of birth, education, etc. Hal, a Hunter, on the other hand, stands far back from the painting and is not interested in the visual details. He also shows little interest in the author's biography. Sometimes interest is increased when one mixes Gatherer and Hunter paintings. One might place colorful abstract paintings (Hunter) in the same room with more staid and realistic paintings (Gatherer) to offer some variety.

The diversity of Gatherers and Hunters.

Here is a simple test that will help identify a Gatherer. George, Helen and Steve are drinking coffee. Bert, Karen and Dave are drinking soda. What does Elizabeth drink? The answer is coffee, because there are two letter e's in her name. A Gatherer might solve this puzzle, but it's less likely the Hunter would come up with the answer or show any interest in such a dry and detailed challenge.

Gatherers are more serious, structured and time-conscious. They like to make plans, work before they play, finish all their projects each day and clear their desks. Hunters, on the other hand, are playful, casual and largely unaware of the time or of being late. They like to "play it by ear" and prefer to play first and work later. Furthermore, they are great at starting projects, but often start too many and are not known for finishing them on time.

When watching a foreign film on television with subtitles, Hal the Hunter is likely to turn up the volume even though he can't understand the spoken language. He wants to get all the benefits of background effects such as footsteps and, of course, the musical score. The Gatherer is happy to read the subtitles without worrying about volume.

Garth, a nine-year-old Gatherer, builds a model airplane carefully, step by step until it is perfect and ready to hang on a wire. Harvey, an eight-year-old Hunter, rushes to complete his model plane and finishes with a few parts missing. Then he lights it on fire and throws it from an outside stairway just to see what will happen and to satisfy his adventurous spirit.

So who's better and who's smarter, the Gatherer or the Hunter? The answer is neither. Gatherers are superior in some areas of knowledge while Hunters are superior in other domains. We might define the true genius as someone who is exceptionally bright in both Gatherer and Hunter modes; someone like Leonardo Da Vinci, who was comfortable designing aircrafts, conducting autopsies and painting magnificent frescoes. Recent research, looking at handedness and brain hemispheric dominance, supports the idea that geniuses may have more symmetrical brain hemispheres, a trait typical in left-handers and the ambidextrous.

Hal and Gail.

Hal is an *extreme* Hunter-type personality. As marketing director for a mid-sized company, he sells mattresses and other bedding products through a dozen retail outlets on the West Coast of Florida. His spouse, Gail, is a *moderate* Gatherer type. She's a lawyer and her personality puts her more in sync with technology and technological attitudes, but she can be creative and intuitive at times. She lives mostly, but not entirely, "inside" the box. Gail worries more than Hal and believes Hal "dampens down" conflicts and needs to be more empathetic and understanding.

Hal's not an anxious person and thinks he has life figured out about as well as one can. He usually sleeps well and may experience some depressive mood swings, but most of the time he doesn't "sweat it."

Hal buys new bottles of shampoo and conditioner if he happens to notice them in the drug store. He believes it doesn't hurt to have extras. Hal starts on a new bottle before finishing the previous one and ends up with a clutter of tubes and bottles in his shower stall. Gail plans her purchases and is careful to use them one at a time.

Hal likes the action and creativity found in the sports and comic sections of his daily newspaper. He thinks most of the other stories and columns are trivial and boring, so he puts them off until evening when he may skim a few pages, looking for two or three articles or opinion pieces that he thinks make sense. Gail feels compelled to read most of the newspaper because she doesn't want to miss any detail that could have an impact on their lives. She starts her day with the obituaries, while he starts his with the sports section and movie reviews.

Gail believes Hal's closet is messy, but he claims to know where everything is, "sort of." He groups his clothes (throws them on hooks or wherever) by concept rather than content. For example, he hangs his dressy short-sleeve and dressy long-sleeve shirts together and hangs his casual short-sleeve and casual long-sleeve shirts together. Gail thinks he should separate his clothes by type, not function. And wouldn't it be lovely if his shirts and ties were sorted by color, as well?

Hal's ties are looped over two hangers and he can pick out a tie for business or social events in less than ten seconds. (This may be a quick if not highly reliable test of whether or not a man falls into the Hunter personality sector.) He may combine plaids and stripes and seldom worries about matching colors, but his overall sense of taste is good. His shoes are usually not polished, but he may shine them if he feels the occasion calls for it.

Hal and Gail disagree about where to keep Hal's cell phone charger. Hal leaves it on a table where he keeps his wallet, keys and other items related to his car and going out. Gail wants him to put it out of sight for appearance's sake and where there is easy access to an electrical outlet. She is relying on logic and a desire for neatness, while Hal is grouping by function.

While speeding along to work in his bright red convertible, Hal misses his usual turn-off point. He seldom looks at street signs to guide him, but he has made that turn a hundred times and can't understand how he could do such a stupid thing. Why did he miss his turn-off point? He didn't realize it was President's Day and some businesses were closed. While Hal was driving along in a semi-conscious, relaxed state listening to a sports station, he relied on sub-conscious cues involving time and distance. With less traffic, he arrived at the turnoff much sooner than expected and missed it. (Now we're starting to see why Hunters have so little in common with machines...unless computers have recently developed subcon-scious mechanisms.) Hal rarely uses his cell phone when he is driving. He's not afraid of an accident; he just enjoys the quiet time and listening to his sports station. His Gatherer bosses stay on his trail and sometimes he takes their calls while driving.

Hal has excellent spatial skills and pilots his car effortlessly. He's adept at picking out the distant silhouettes of police cruisers through his rear-view mirror and the subtle brake-tapping of cars hundreds of yards ahead, where a State Trooper lies in wait. It irritates Gail that he has never been cited for speeding, even though he drives well over the speed limit. Deep down, Gail really hopes he gets a ticket. She rationalizes that it would curtail his speeding and protect him from injury. When she's honest with herself, she has to admit that sometimes his approach to life drives her crazy and she'd like to shake him up and bring him down to the earth she lives on.

She even fantasizes about it some nights when she can't sleep. Man, what would he say if he got a big ticket? He'd probably hide it, but she'd find out sooner or later. That would take some of the swagger out of him. But lo and behold, defying all logic and what she knows is good and true, Hal just rumbles along, like some alien spirit, ungrounded by culture or logic.

On the radio, a color commentator and an announcer field calls from listeners. Hal loves to hear the comments of the sports guy, who is folksy and obviously not well-educated. What he says is a welcome homily meant to warm the hearts of athletes and former athletes: "I'm still in remembrance of the way he come up from college ball. Compensation-wise, he has that low stride and demeanor that don't let him get gobbled up by them big defensive tackles." *Ahh, poetry,* Hal thinks.

The polished announcer is somewhat irritating because he talks too fast and mostly about useless details. But there is something relaxing about hearing the two of them conversing together. They generate a low level of pleasurable tension because of their differing personalities and they give Hal a sense of wholeness. With a flash of insight, he realizes they are complementary parts of one person.

Office antics.

Gary is the bookkeeper at Rudy's company. He's an *extreme* Gatherer. Creative thoughts make him downright nervous. If ideas don't have a sequential origin they are mysterious and best left alone. Gary is into logic and control. He has little patience for emotion or what he considers "soft" reasoning. He is waiting anxiously when Hal arrives for work. Gary seems like a nice enough guy, but Hal has serious problems communicating with him. Another irritating thing about Gary is that he stands only ten to twelve inches from Hal's face when they are conversing. *Why can't he respect my personal space?* Hal wonders. He doesn't realize that "robots" have difficulties orienting themselves in space.

Gary has a complaint. Each month Hal sends out a notice reminding the staff that the monthly marketing meeting is in one week. Gary shows Hal the notice he received, dated September 12, that indicated the meeting would fall on August 19, an obvious oversight. *How could that nitwit make an issue out of a harmless mistake?* Hal wonders. Hal is convinced Gary isn't too bright and Gary is convinced Hal is lazy and intellectually challenged.

Hal mentions that his wife will be age forty on her next birthday. Gary says he knows and that he, Gary, will always be a year younger than Hal's wife. Her birthday is October 14 and Gary's birthday is the following January, three months later. Hal shakes his head and stares at Gary in disbelief.

Besides, Hal best understands problems through a kind of osmosis. He enjoys reading fiction about World War II and is reminded of "Emily," a character in Ian McEwen's novel *Atonement*: "Many hours of lying still on her bed had distilled from this sensitivity a sixth sense, a tentacular awareness that reached out from the dimness and moved through the house, unseen and all-knowing. Only the truth came back to haunt her, because of what she felt, she knew. The indistinct murmur of voices heard through a carpeted floor surpassed in clarity a typed-up transcript; a conversation that penetrated a wall or, better, two walls, *came stripped of all but its essential twists and nuances*....The less she was able to do, the more she was aware."[10] (My italics) That's *my kind of gal*, Hal murmurs.

Hal's brain is reminiscent of a character from Thomas Harris' novel *Red Dragon*: "[He] had a lot of trouble with taste. Often his thoughts were not tasty. *There were no effective partitions in his mind*. What he saw and learned touched everything else he knew.... [He was sorry] that in the bone arena of his skull there were no forts for what he loved. His associations came at the speed of light. His value judgments were at the pace of a responsive reading. They could never keep up and direct his thinking."[11]

Later, Hal feels hungry and announces he's going to lunch. Gary glances at the clock and says, "It must be time to eat." Hal wonders why Gary always looks at the clock before deciding to eat. Doesn't he know when he's hungry? Gary offers Hal a ride to lunch. Hal is reluctant, but Gary helped him out of a couple of tight jams recently, and Hal feels he owes the poor guy.

Just the week before, Gary had been happily absorbed in an activity he finds truly relaxing and therapeutic: examining checkbooks. During his casual review of the sequentially indexed checks, Gary discovered a check was missing from the back of a marketing department checkbook and a fraudulent check had been written. Hal had bent company policy by allowing a pretty new secretary to write checks. Hal had one of his many *Blink* moments when he relies on gut-level feelings and felt he could trust her by "just looking at her." Gary had saved him from embarrassment and a possible reprimand.

Gary putters along carefully in his white, polished, high-mileage sedan. Hal had noticed that no stickers adorn the back of Gary's car except for an Automobile Association tag. Gary is in no hurry and chatters in an unemotional, monotone voice about the weather and office gossip. Gary gives Hal a litany of reasons why he believes a recent management decision was incorrect. He traces at least twelve logical steps to arrive at his conclusion. Hal finds he's both irritated and sleepy. Gary may be correct, but the decision wasn't a big deal anyway, in light of the company's overall goals.

Gary is a conservative driver, but not a good one. He stays too close to the cars in front of him and when they slow down, waits until he is almost too close before slamming on his brakes. In the restaurant parking lot, Gary tries to park his car between the designated parking lines but finally comes to a stop with the right front tire over the diagonal parking stripe. Hal prides himself on his perfect diagonal and parallel parking. He never lets his tires even kiss a diagonal parking stripe. Much to his wife's dismay, if an adjacent car parked too close to the line, Hal has no compunction about ignoring the lines altogether and parking his car at an equal distance between adjacent cars, even if he has to park well outside of one line.

After lunch, Gary turns into a church parking lot to drop off some magazines for recycling and pulls into line behind three other cars. Hal notices that the first bin is for newspapers and the magazine bin is well beyond the newspaper bin. He suggests that Gary pull around the other cars. "We'd better stay right here in line," Gary intones.

After work, Hal notices that one of his tires is a little flat and pulls into a tire center. An unshaven and disheveled mechanic comes over to the car, and Hal points at the left front tire. The mechanic bends and peers at the tire and then jerks his head toward a nearby hoist. Hal drives over and rolls carefully onto the hoist. He gets out and watches the mechanic pull the tire and test it. After patching the tire, the mechanic scribbles something unreadable on a ragged piece of greasy paper, gives it to Hal and motions toward the office at the front of the center.

Hal nods and ambles over to the office where he finds two clerks at the front counter working behind their computers. They look at him, but their faces are emotionally flat and signal no recognition. Finally, a young woman with a bright green skirt, lots of makeup and a winning smile bounces along the corridor from a back room and gives him a friendly wave. Hal smiles back and hands her the bill, along with his credit card. She swipes the card and Hal signs off. With a personable bow and a "thanks," the woman gives Hal his receipt and he's on his way.

He feels really good about the tire center and decides to use it again in the future. What he doesn't realize is that much of his positive reaction is a result of only one word being uttered during the entire analysis, repair and payment for his tire. Hal prefers nonverbal communications to verbal ones, although he has learned to use canned talks for sales presentations.

Sometimes Hal's verbal reticence gets him in trouble. During one tense marital period he and Gail had a huge argument and were giving each other the silent treatment. Hal realized he had to catch an early flight the next morning, and he usually counted on Gail to wake him up. He didn't want to give in, so he wrote Gail a note asking her to wake him at 5:00 A.M. The next morning he awakened at 8:30 AM and found a note from Gail. "It is 5:00 A.M. Wake up."

When Hal gets home, it's dark and a clock repairman is talking to Gail about their grandfather clock, which had stopped and which the repairman had reset. He is explaining a complicated formula for determining the phase of the moon in order to reset the moon dial on the face of the clock. Hal had noticed the half-full moon on his ride home from work and could now see the moon through a nearby window. Without comment, Hal reaches up and turns the dial to a half-moon position. "What are you doing?" Gail crosses her arms and fumes. Hal glances out the window at the moon and then at the repairman.

"Won't this work?" he asks.

"Well, I guess so," replies the repairman, "but it's probably not totally exact."

Extreme Gatherers are susceptible to the mechanization culture. They already are robot-like and lacking emotional responses. But the greatest loss is the Hunter, who demonstrates those sentimental, sensibility and sensitivity qualities.

3

Time to Inoculate?

Electronic Impact on Kids

Kids and learning.

In 1952, a 32,000 page compilation of a fifty-four volume set called *Great Books of the Western World* was published. At the time, these selected works included those of Aristotle, Milton and Locke. According to the article "Lessons From the Great Books Generation" by L. Gordon Crovitz, "By the 1980s, university students objected to the very idea of the Western canon, chanting 'Hey, hey! Ho, ho! Western culture's got to go!'"[1] Today, some young people believe their trial and error approaches to learning, fueled by electronic games and the Internet, are all they need. But, in spite of proclaimed maturity and distrust of authority, some seem more dependent than ever and many are postponing adult responsibilities for years. A growing number, according to Crovitz, have moved back home with their parents.

Go with the flow.

Douglas Rushkoff's *Playing the Future: How Kids' Culture Can Teach Us to Thrive in An Age of Chaos* tells us that kids relish a technological climate of chaos and go with the evolutionary flow, while their elders cling to obsolete institutions.[2] Business executive Don Tapscott, in his book *Grown Up Digital: How the Net Generation is Changing Your World,* concludes that Net Geners are smarter and quicker than their predecessors.[3] Is opposition to the machine world really generational? And if so, is that positive or negative? Is this a matter of matching the older generation against the younger techies?

Imagine this scenario: A gray-haired professor totters to his chair in the corner of the ring. He's a small man and his boxing gloves seem to weigh him down. About a dozen trainers and advisors sit in his corner, just outside the ropes. They're old-timers, too.

Meanwhile, a young techie climbs through the ropes and whizzes about the ring on an electronic skateboard, while juggling a laptop computer and a cell phone. His jeans are low in the back, his colorful underwear flashing under the glaring spotlights. Books, artwork and flowing green plants litter the professor's corner while the young techie's corner is jammed with flat screen televisions and electronic equipment. The techie is alone in his corner, but several hundred of his young advisors sit just outside the ring, texting on their cell phones and staring at their laptop monitors.

The bell for the first round rings and the techie hurdles out of his corner. He moves quickly around the ring but is frequently distracted with instant messages from his advisors. The techie's bearing and broad smile denotes supreme confidence. He's sure the old man is no match for him. After all, his sources of strategy and information are limitless. Insightful and creative suggestions are sure to follow. He scoffs with disdain at the old professor's tiny group of advisors.

The young techie cartwheels, somersaults and backflips his way around the old professor, but seems confused with all the advice he's receiving. He lands multiple blows, but none of them are substantial or punishing. In the ninth round of this ten round match, he is clearly ahead on points. The professor leans back against the ropes. This confuses the young techie who, while spinning around the ring at a fast rate, seems to be swinging wildly and at random.

The professor reports to his trainers that the young techie never looks him in the eye. His vision follows the ring's periphery. But at this rate, the professor knows he'll lose the match. He turns to his advisors and insists they give him some winning advice. Several of his trainers page through ponderous books and begin to read.

"Hey guys," the professor says enthusiastically. "I need help now. This can't wait."

"Be patient, my son," his trainers say. "Stay steadfast and true."

The bell for the tenth and final round rings. The young techie charges out of his corner swinging in every direction. The professor turns to his trainers:

"Well?"

"Okay," they respond. "Our wisdom comes from the classics. Because his eyes follow the horizon instead of the heart, he has over-trained his peripheral vision and won't see a blow coming from below. This is our advice: Lie back on the ropes and when he comes past give him an uppercut. Be patient, my son, and wait for the proper moment. He's forgotten the ultimate goal of the match...which is to win."

He knocks the techie out with one punch and the fight is over.

If you doubt that the old professor could defeat the young techie, look at the confrontation between computers and chess masters. Charles Krauthammer, syndicated columnist with the Washington Post Writers Group, reported that Russian chess champion Garry Kasparov took on the X3-D, Fritz, a computer program.[4] Being a computer, Fritz has no imagination. Fritz could see twenty moves deep, but even that impressive knowledge yielded no original plan of action. Kasparov, on the other hand, came up with a strategic plan to push a single pawn down the flank to queen. Meanwhile, Fritz was reduced to shifting pieces back and forth. At one point, Fritz moved his bishop one square and then back again on his next move. No human would do that. Not just because it's a waste of two moves, but because it's simply too humiliating. "This move showed that the computer doesn't feel any embarrassment," said another grandmaster.[5] Kasparov won.

When Facebook.com was opened to adolescents, college students didn't like it one bit. I imagine that when our current adolescents reach college age they won't want young individuals sharing the goodies with them, either. But by that time, even elementary school students will be invading their privacy. It's similar to letting a younger brother or sister share the same bedtime or allowance.

Older adolescents and college students believe they have somehow earned the privilege of immediate gratification through machinery and think younger people should work for it. But don't the young techies believe we're all at the same level, despite age or accumulated book learning? Maybe they like "leveling up" better than "leveling down."

The new order hasn't prevailed quite yet. Even those putting together Wikipedia, the online encyclopedia, have had second thoughts about the "wisdom of the crowds." Because some entries proved to be unreliable, new guidelines for adding entries are now required to provide reliable sources. L. Gordon Crovitz writes in *The Wall Street Journal* that "the guide credits old media and old-fashioned definitions to establish legitimacy."[6] This means some editorial control, just like in the old days. In the past, enterprises usually started at a high professional level and then fell to the "wisdom of the crowds." Music is an example. With IT, it's the other way around. People start from the bottom, thinking they're already at the top. As they get a little wiser, some finally ask for help from individuals who know better.

Man can't live on cheese alone.

Did we really start learning in preschool or earlier, when our parents read to us? Did the light really come on in grade school or middle school or as late as high school? For some of us, it wasn't until college or even graduate school when we seemed to get the hang of it. And for some, it was a mid-life change in their thirties or forties.

The necessary conditions for learning include internal motivation, internal and external rewards and prioritizing. Forgetting—or at least not paying attention to—some stimuli is as important as getting the information into our brains so that we can remember it. The skill of prioritizing works its way into a broader concept, that of proportionality. Some video games consist mainly of chasing, shooting or running from someone. This might build more brain cells for speedy responses under meaningless visual challenges, but what mental age is required?

Looking at developmental norms like those on the Vineland Social Maturity Scale and the Stanford-Binet Intelligence Scale, as found in *Tests and Assessment* by W. Bruce Walsh of the Department of Ecology and Evolutionary Biology at the University of Arizona, and Nancy Betz, Department of Psychology of Ohio State University,

I would put this challenge at a mental age of three to six years. Even before the age of three, kids are expected to follow instructions of the "if-then" form, and at age four, they can follow school or facility rules. Before the age of five they build three dimensional structures, complete inset puzzles of at least six pieces and unlock key locks.[7]

Some basic electronic games are reminiscent of rats learning in a maze. In such experiments, the rats hurry down to the end of the runway to reach the vertical bar. They use their little pink noses to give the toggle bar two pushes to the right followed by one push to the left. And finally, the box containing the yummy cheese opens up.

After finally mastering that sequence, the rats find that the combination no longer works. Now they must give two pushes to the left and one to the right to open the box and receive the reward. Basic electronic games condition our children in a similar manner.

Our video games lack *density*, which is a measure of progress over time. They take a lot of time to complete and that's because the goal of the game designer isn't learning—it's entertainment. Keep the rats scampering and don't run out of cheese.

The games, at this level, give players quickness at the expense of wisdom, detail at the expense of context and proportionality, and reinforcement of reflexive learning at the expense of abstract thinking. Much of this learning engages the posterior (toward the back) parts of the brain, which records information (the junk in junk out area). It may also engage non-abstract areas of the frontal lobes such as motor expression (e.g. writing).

So, these basic games leave out the important area of the frontal lobes. The analogy to an orchestra conductor is often cited. The orchestra leader directs musicians from different and competing areas of the orchestra. In a similar fashion, the frontal lobes of the brain take disparate chunks of information that are necessary, but not sufficient, for high-level thinking and combines, collates and prioritizes that information to offer an integrated and unified expression just like the orchestra leader.

Tired of Twittering?

There are many definitions of neurosis and other mental conditions that can significantly impair one's ability to function interpersonally.

After many years in private practice, it became evident to me that one of the most predominant features of neurotic dysfunction is the tendency to over-focus on one thing without balancing that idea, fact or event with other facts and ideas to permit a more nuanced picture. The neurotic person lacks context, relevance and proportionality. It's like wearing blinders and not seeing—or seeing too much of too little. There can be many reasons for the etiology of a neurosis and people may argue about the causes, but in the end many neurotic people demonstrate this feature.

Will game players become neurotics because of exposure to electronic games? No. But excessive use of basic games at early ages can reward and encourage a child to think narrowly and even shape brain cells in the direction of detail and over-focus.

Are you part of the Twitter "revolution"? This is a social networking application for short text messages of no more than 140 characters. Users log into electronic diaries and keep their "friends" updated with every minute-by-minute detail of their lives. This exemplifies over-focusing and thinking narrowly. Biz Stone, one of the Twitter founders, was at a tech conference where he observed groups of people Twittering each other. As reported by Michael S. Malone in *The Wall Street Journal*, "...it was like seeing a flock of birds in motion."[8]

Like the 1963 Alfred Hitchcock suspense classic, birds are everywhere, diving and circling without destination or purpose. This is reminiscent of the mythical "Koo-Koo" bird, which only flies backwards and doesn't care where it's going; it only wants to know where it's been. This is a great way to distract a person from oneself, which may increase one's sense of self-importance and narcissism. I'm reminded of a question on the Minnesota Multiphasic Personality Inventory (MMPI), one of the most widely used personality tests. In the 1960s, answering yes to the statement "I am a very important person" placed one in the narcissistic range and perhaps even the delusional range. By today's standards, the same answer is viewed as normal.

Of the two personality types we've discussed, Hunters and Gatherers, the latter are closer to the computer in terms of their learning style and even their personality makeup. They have an intuitive knowledge of the computer's logical, precise and sequential operations. This may encourage them to be one-dimensional rather than balancing their personalities with the positive traits of their Hunter sides.

The Hunter, on the other hand, may resist the dry, precise and sequential aspects of the machine, but will find the adventurous electronic games highly stimulating. If the Hunter accommodates to IT thinking at an early age, he or she may lose creativity, spontaneity and artistic inclination. The Hunter's language may become bland and he or she may take fewer entrepreneurial risks.

Monkeys and robots.

Some people believe that the computer is equal to the brain or will eventually become equal to the brain in its ability to think, reason and even emote. They compare transistors and computers to the synapses in the brain, but what they forget is that brain synapses involve chemical as well as electrical energy. In a *Wall Street Journal* article by Lee Gomes, Christofer Koch, a scientist at Cal Tech, warns that increasing the number of transistors will not have a brain-like effect. He states, "With bigger computers, all we are going to get is more junk. It would be like someone in 1900 saying, 'Give us more slide rules and we will understand the universe.'"[9]

While we know that the computer can't operate like the human brain, we also know that the computer and any other learning system can change the brain, especially during younger years when the brain is more plastic and open to change. In the article "How Thinking Can Change the Brain" Sharon Begley claims that "scientists at the University of California, San Francisco rigged up a device to tap monkeys' fingers 100 minutes a day every day." While the monkeys were tapping their fingers they heard sounds through headphones. Some of the monkeys were taught to ignore the sounds and pay attention to what they were feeling, while others were rewarded for paying attention to the sounds.[10]

The article continued to say, "After six weeks, the scientists compared the monkeys' brains" and found that when monkeys "paid attention to the taps" that area of the brain doubled or tripled, but when the monkeys paid attention to the sounds there was no expansion. And the area compromising the auditory region increased.[11] It doesn't take much repetition in a monkey's brain to alter it. The Jesuits, an order of Catholic priests, have a saying: "As the twig is bent, so the tree shall grow." This refers primarily to the first six years of human life. Maybe now the vulnerable little twigs are getting a good bending and hopefully, not a good twisting.

One eight-year-old I observed was sitting about six inches from a television screen, manipulating an object with his thumbs. When I asked him if he was playing one of his electronic games, he said he was just watching television. In fact, he was watching a movie while pressing the volume keys up and down on his television remote with his thumbs. Where did he learn this? Was it while finger painting or playing football in the backyard? I doubt it.

Our primary concern here is not with the content of computer games but rather the focus or process. Young children are learning *something* when they sit down with computer games, text message or speak on their telephones. As yet, we don't know what effect this is having on their brains' neural pathways and circuitry; we certainly don't want to lose some of the human qualities that are exemplified in the Hunter personality.

These include flashes, shifts, epiphanies, intuitive thinking, non-verbal responses, visual-spatial skills and the ability to synthesize; not just analyze. We don't want to give up the holistic world to coding and segments, nor do we want to give up the ability to prioritize and bring proportionality into our decision making. Logic and content have their place and, without the Gatherer personality *and* computers, we could not function in today's society. But to achieve our highest level as humans, we need emotion, sensitivity and creativity.

How children learn.

Rather than starting with machines and machine learning, let's take a look at how children learn. The first ingredient is curiosity. I believe that curiosity is an inborn and *internal* motivation that does not need much prompting from the outside. Harry Harlow, psychologist at the University of Wisconsin, established in the 1960s that monkeys would rather open a door to watch a model train run around a track than eat. So curiosity is a basic drive.

Since we know that motivation is an innate and basic drive for humans as well, children don't have to be driven by a stimulus from outside in order to learn. Sure, we can use an external reward from time to time, whether that's candy, a smile or a pat on the back. But human learning is largely self-reinforcing and accomplished gradually, from within. The child acquires certain skills, takes pride in accomplishing those skills and sees them as stepping stones to satisfy his or her curiosity and achieve higher levels of knowledge.

However, when outside rewards are piled on too quickly in response to modest efforts, they tend to extinguish our desire for true learning. Basic electronic games offer inordinate external rewards. The child learns *something*, but the probability of those rudimentary skills generalizing to important learning processes is low. We know from experience and studies with children that giving them money for homework does not increase learning in a sustained manner. What the child is really learning is how to acquire money, tokens or candy, not to satisfy an inner thirst for knowledge.

Video games are not designed to teach—they are designed to entertain. To their credit or discredit, they do an impressive, almost scary job of entertaining. Testing children who are hyperactive or disruptive may require the use of immediate external rewards such as candy for each of the child's responses, but this is done as a last resort in order to try to dig through disruptive behavior to discover the child's potential. It is not recommended for teaching most kids on a sustained basis.

With computer games, the child is exposed to dazzling graphics. This provides a breathtaking external motivator at the expense of self-directed, internal motivation (the reward or motivator comes from the outside, not from our thinking and activity). The reinforcement (reward) schedule also differs with computer games. The child can count on a 100 percent response to his action if it is performed correctly. As documented by authors Douglas A. Bernstein and Peggy W. Nash in *Essentials of Psychology*, psychologists call this a continuous reinforcement schedule and it results in a fast response rate.[12] This isn't true in the real world. Most people, especially extreme Hunters, are a little unpredictable. We're never sure how they'll respond to us.

Unlike their cold-blooded cousins, the machines, people throw more curve balls than straight balls. In the real world we can follow the rules and make the correct responses, but we never know for sure if we'll be rewarded for all that hard work. Psychologists call this a variable-interval schedule of reinforcement. It often results in a slower response rate than continuous reinforcement schedules. Let's say we're trying to convince a purchasing agent to buy our product. We do a great job and expect to make the sale, but we remind her of the red-headed bully in third grade who ate her lunch and called her Dumbo or maybe she already promised to buy from a friend who takes her to lunch a lot.

This kind of reinforcement in the real world actually results in learning that lasts longer and is harder to extinguish than reinforcement 100 percent of the time. If you're at the county fair and receive a dancing Elvis figure every time you throw a ball in a box, you will soon know what to expect and you lose some of your drive to succeed, not to mention alertness and vigilance. Boredom might even set in. Intermittent rewards tease your motivational system, keep you on your toes and generalize to real-world situations where rewards are also intermittent.

Good learning.

Maybe good learning is like making good wine. Remember the old saying: "No wine before it's time"? What happens when we harvest the wine before it's ready? It's quicker and requires less patience on our part...and the first sip of wine isn't half bad—impressive, even. But depth and complexity are missing. It's just table wine that won't improve over time. The most enjoyment is in the first sip. After that, it's all downhill.

Growing grapes is not easy. If the seeds are planted in shallow soil, they may produce for one or two seasons but soon the roots will die. Growing quality grapes requires the right seed and the right soil, along with patience and knowledgeable mentoring. When fine grapes are fully mature, combined with other fine grapes and placed in a wooden cask to breathe and ferment, we eventually have a quality wine of deep color and pleasant nose. And this living wine will continue to mature, even inside a glass bottle in the cellar.

Wouldn't it be nice if we could plant grapes one week, harvest them the next week and enjoy their superb flavor with our foods only a few days later? Unfortunately, wine, like learning, requires hard work and patience. Even then, the results are not guaranteed.

Facts and thinking.

IT learning gives us lots of facts, sometimes more than we want or need. In *Stumbling on Happiness* by Daniel Gilbert, a study was designed to show that extra explanation and giving additional facts may rob the event of its emotional impact. Researchers gave college students one of two cards with a dollar coin attached and then walked away. Both cards stated that the researcher was a member of the "Smile Society," which was devoted to "random acts of kindness." But one card also contained two extra phrases: "Who are we?" And "Why do we do this?" These phrases did not provide new information, but they may have made the students feel as though the event had been somewhat explained.

Later the researchers surveyed the students to find out how they felt. Those students who had received a card with the explanation felt less happy than those who received a card without it. The same research also showed that despite feeling better without the information, all the students polled said they prefer having all of the information whenever possible.[13] Aren't we humans strange critters? Makes you wonder what kind of information we're getting when we ask kids to tell us about their new world and new learning styles.

Gatherers and Hunters learn differently; Gatherers move rapidly through sequences and build conclusions from the bottom up. They might put a puzzle together by matching the outlines of each piece. The Hunter, on the other hand, is more likely to look from the top down at the entire puzzle and put the pieces together based on a perception of the whole.

Another difference between Gatherers and Hunters is music appreciation. *The Journal of Research in Music Education* indicates that right-brain individuals (Hunters) seem to do better in visual conditions whereas left-brainers (Gatherers) do better in verbal conditions.[14] Dennis Prager, a notable radio talk show host, has reported that he can listen to words, dates or political facts and repeat them almost verbatim three weeks later, but cannot recall the words to the Beatles song "Hello Goodbye".[15] I believe the music and rhythm may disrupt his memory of these simple words or they're not getting into storage to begin with.

According to an article in *Neuropsychology* by Guy Vingerhoets, researchers have found that when adults listen to *what* is said, blood flow increases significantly on the left side of the brain. When participants shift attention to *how* something is said—tone of voice, whether happy, sad or anxious or angry—blood flow goes up markedly on the right side of the brain.[16] The Gatherer shares some traits with the so-called left-brained individual and the Hunter shares some traits with the so-called right-brained individual.

When some students are engaged in computer games, they learn to improve visual alertness, especially in the visual periphery, and to move laterally from left to right and right to left (directionality). But if they are playing basic data games that are rote and repetitious, they may be operating in a semi-conscious stupor. In such cases students may not be learning, because they're operating on autopilot. We have all experienced walking or driving a familiar route and arriving at our destinations with almost no memory of our trips.

According to *Gut Feelings: The Intelligence of the Unconscious* by Gerd Gigeirenzer, individuals who have brilliant memories for facts sometimes have difficult times thinking on abstract levels. They may find it difficult to recognize faces and while they are excellent word callers when reading, do not understand the context of what they have read and cannot summarize the substance of the story.[17]

Busy work.

We are not just discussing learning content through electronic systems, but rather how these systems may change the brain and gradually and insidiously alter the way we approach learning and other people. In Mizuko Ito's interviews with children, some kids, even though they had enjoyed "the game" over the years, admitted the hold it had on them and indicated they had lost valuable time and could have been more productive in school. Some also see their game play as a waste of time and as killing time.[18] Hopefully, that's all that's happening to their brains.

To better understand where IT may be taking us, let's visit Tom Sawyer, a fictional character in Mark Twain's classic book, *The Adventures of Tom Sawyer*. A street-smart master of non-verbal communications, Tom is truly a Hunter. He doesn't like school and he's not crazy about boring, repetitious work. When Aunt Polly makes him whitewash the picket fence in front of their home, Tom develops a plan. He starts whistling and paints as though he's really enjoying it. Before long, some of the neighborhood boys come around and ask if they can help. He's reluctant at first, playing a little reverse psychology. Finally, he relents and lets them help, but charges them money for the privilege. Tom sits in the shade, eating an apple and enjoying the other valuables he has received. "If he hadn't run out of whitewash, he would have bankrupted every boy in the village."[19] Here we have a socially aware person who figuratively has green grass growing between his toes. His real world entrepreneurial scheme is a great success.

Fast-forward a hundred or so years to thirteen-year-old Garrison, a Gatherer, who is the only son of sophisticated professionals. Garrison lives in a high-rise condo in Manhattan and spends tons of time behind his computer playing electronic games and visiting Web sites. His toes have yet to meet green grass and he doesn't interact socially, except for some ghostly "friends" on Facebook. He distracts himself through instant messaging and multitasking. His parents are proud of his "creativity." He got an "A" for a book report on Mark Twain's *Adventures of Tom Sawyer*. Garrison had researched online some literary and historical sites, cut and pasted—and presto! Another quick production that still left time for video games.

His creative efforts are uninspired compared to Tom, the uneducated country bumpkin who comes from a rural, lower socioeconomic background. As reported earlier, Ito's study shows that some low income parents encourage their children to engage in real world activities and can't afford many of the high-tech gadgets available to middle and upper socioeconomic kids.

Multitasking: Remedy or shell game?

We hear a lot about the benefits of multitasking (the performance of multiple tasks at one time) today, both from young children and adults. Handling input from many sources simultaneously should keep us tuned up, alert to our environment and lead to an even greater capacity to learn. While kids *are* learning facts and some visual acuity, aren't they also learning to multitask?

A study reported in the *Journal of Experimental Psychology* entitled "Is Multitasking More Efficient?" showed that multitasking may double the time it takes to do tasks compared with doing them one at a time. David Meyer, a psychologist at the University of Michigan, states that there is little evidence youngsters exposed to IT are any better than adults in acquiring knowledge. "In four experiments, young adults switched between different tasks." The results showed that for all types of tasks, subjects lost time when they had to switch from one task to another and time costs increased with the task's difficulty. Meyer points out that a mere half second of time lost to task switching can mean the difference between life and death for a driver using a cell phone, because during this time the car is not totally under control and it can travel far enough to crash into obstacles that otherwise could have been avoided.[20]

We would expect Net Geners (younger individuals exposed to IT learning) to be more familiar with communication technologies than people ages thirty-five to thirty-nine. Because of their multitasking experience, they should also handle interruptions better than older men and women. Right? According to the study "Cognitive Constraints on Multimedia Learning," published in the *Journal of Educational Psychology*, eighteen to twenty-one-year-olds performed 10 percent better on intensive problem-solving exercises without disruption than the thirty-five to thirty-nine-year-olds. However, interruptions such as phone calls and text messaging caused the Net Geners to lose their cognitive advantage over their older counterparts.[21] So even while the IT youngsters were better able to use the new technologies, they were less effective at recovering from interruptions! The interruptions did not significantly change the performance of the older generations.

Recently, I observed a six-year-old and her friend. They were sitting at an outside table at a café. It was an informal place with plastic spoons and paper cups. Their mothers glanced at them nervously from time to time, hoping they wouldn't spill something. Naturally, one of them knocked over her cup of milk. This is not unusual for a six-year-old. However, the explanation is found in multitasking.

The little girl was talking to her friend, listening to a small radio and watching her parents, the waiters and other customers. Are we born with multitasking capacities? As the brain matures and we learn from parents and teachers, we begin to inhibit the temptation to multitask. We begin to focus and prioritize. Perhaps we don't need to learn multitasking. Perhaps we need to *unlearn* multitasking. I have to wonder if the multitasking label isn't an attempt to make a virtue out of distractibility and shallowness.

Author Walter Kirn states, "This is the great irony of multitasking—that its overall goal, getting more done in less time, turns out to be chimerical. In reality, multitasking slows our thinking. It forces us to chop competing tasks into pieces, set them in different piles, then hunt for the pile we're interested in, pick up its pieces, review the rules for putting the pieces back together, and then attempt to do so, often quite awkwardly."[22]

Adding on-screen text can overload the visual information processing channel, causing learners to split their visual attention between two sources. In four experiments, college students viewed an animation and listened to concurrent narration. When students also received concurrent on-screen text that summarized or duplicated the narration, they performed worse on tests of retention and transfer than did students who received no on-screen text. Lower performance also occurred when authors added interesting but irrelevant details to the narration or inserted interesting but conceptually irrelevant video clips.

In one example a federal jury watched the video of a drug deal occuring. The camera, hidden in the car's front vent, took decent footage, but the soundtrack was poor. To assist the jurors, the court deciphered the conversation and added words to the screen. After viewing the video, the jury was ready to acquit, because they hadn't seen any drugs. The jury foreman, a Hunter, had ignored the text and kept his eyes on the picture. Much to the surprise of his fellow Gatherer jurors, he said he saw the drugs and urged a guilty vote. After some discussion, the judge allowed the jurors' to watch the video again. This time they ignored the on-screen text. Ten minutes later they found the drug dealers guilty as charged.

Cyber kids.

Ask any grade school teacher. One of the problems with Internet learning is what the student might *not* be seeing. One cannot learn what one is not exposed to, and without some structure and guidance, students can waste enormous amounts of time. Students may use the new search technologies to investigate only those areas *they* think are relevant, but may not have the background or knowledge to know what is truly relevant or potentially relevant.

Despite alternative ways to get the news, the demise of the daily newspaper may be quite damaging. When people read newspapers they learn "newsworthy" information that catches their interests and has relevance to their lives. Relying entirely on surfing the Web is narrow and egocentric. It's easier and quicker to scan printed material that has been organized into sections than to labor sequentially through one blurb after another. Today, many adults eschew newspapers and rely entirely on the Internet. They may find something they're interested in, but meanwhile, they could miss important connections to their occupation, education or family.

Does this bombardment of visual and auditory stimulation enrich our thinking or does it distract us from what is wise and true? Psychologists in training are taught to look behind words and listen with their "third ear." One is reminded of the song "Razzle Dazzle" in the Broadway musical *Chicago*. What's behind all that noise and clutter? I believe it's often a way to manipulate and hide the truth.

In order to make a real impact, perhaps the world of silence is more important. When the late Pope John Paul II visited the Holocaust Memorial in Jerusalem, he called for silence. At the Wailing Wall, the pope said, "Some things are beyond words and before words." The Gettysburg Address was short on clutter, as well.

Some New York City pubs are advertising quiet time. They are sans television sets and raucous music. The silence can be unnerving. This leaves only self-initiated entertainment. What to do? Sit and think? Let the mind wander? Talk to a friend? Really communicate?

I'm reminded of what clinicians used to call the "Sunday Neurosis." People would stay busy during the week, but had time to think about their lives when Sunday rolled around. When I was interviewing for an internship position at the University of Wisconsin, the psychologist put me through a stress interview. What did he do? He said absolutely nothing. He sat there, unspeaking, and I sat there nervously wondering what the heck to do. Will our cyber kids be able to handle the stress imposed by silence? I doubt it.

Calling all kids.

Some believe the IT environment hasn't hurt student learning and, in fact, has helped it. The United States Department of Education reported on December 10, 2008, that U.S. students are making gains in mathematics scores.[23] We should note that two of the states had much higher scores; in those two states, half of their schools failed to meet annual progress goals. The parents in Massachusetts and Minnesota are wealthier and more highly educated than U.S. averages. Furthermore, critics connect these rising scores to improvements in math and other instruction that has its roots in state programs starting in the 1980s. Additionally, seventeen-, eighteen- and nineteen-year-olds were not exposed to the IT environment in its present form until they were already ten years of age.

Students now use calculators even when taking the Scholastic Aptitude Test (SAT). As adults, these youngsters may not have the basic math skills they need to make quick decisions in face-to-face situations. Reading literacy also seems to have taken a hit. The National Endowment for the Arts surveyed voluntary reading, not reading required for work or school. Between 1982 and 2002, eighteen- to twenty-four-year-olds dropped from a literacy rate of 60 percent to 43 percent. This can be compared with forty-five-year-olds and above where the drop was only two or three percentage points.[24]

Mark Bauerlein, author of *The Dumbest Generation: How the Digital Age Stupefies Young Americans and Jeopardizes Our Future*, presented data from the Nielsen Norman Group that shows how Internet users read online. Nielsen Norman needed accurate information in order to satisfy the needs of major companies and institutions. The company's work is practical, hands-on and longitudinal, extending over several years. User trials included eye-tracking components that detect eye movements on the Web page and measure moves and rests. Nielsen Norman concluded, after fifteen years of assessment, that Web users don't read very well. Only 16 percent of the subjects read text on various pages linearly, word by word and sentence by sentence. The rest scanned the pages, choosing individual words and sentences out of sequence. Nielsen Norman believes the screen encourages viewers to focus on items that interest them and pass over the rest.

In 2005, Nielsen Norman Group focused on teenagers and the group's study concluded that the common belief that teenagers wield technology better than their elders may not be the case. "Overall, teens displayed reading skills, research procedures, and patience levels insufficient to navigate the Web effectively."[25] Nielsen believes these incorrect stereotypes result from self-reported behavior of teenagers rather than close observation of their actual behavior.[26]

Robotic armor.

How does one start out human and end up machine-like? It's easy, especially for Gatherers like Gail, Gary and Ginger. Parents, teachers and peer groups urge children forward and reward them for taking on robotic armor. And, of course, the machines themselves dole out immediate and tasty rewards, like cheese bits to rats in a research lab. Children are gradually shaped and become adults.

There are many positives in today's digital culture, but we need to examine electronics' impact on all tried and true psychological variables that influence our lives, among them: emotion, stress, repression, separation, guilt, shame, addiction, sexuality, aggression, identity, trust and self-disclosure.

4

Twittering Our Creativity Away

Creativity and Computers

The Hunter personality type is usually more creative than the Gatherer. Creativity is considered by some to be a first place position in the world's economy and is promoted as the unique characteristic that separates Americans from the more staid and structured citizens of other societies that, according to tradition, may be Gatherer-dominated.

Creative.

Psychologists do not agree on a single definition of creativity. Some view it as an innate ability and others believe it's a conscious decision to go one's own way.[1] The essence of creativity is primarily a Hunter prerogative, but recently psychologists have added the requirement of *successful production* to their definition of creativity. According to this definition, one must not only show creative insight, but must also develop that insight to a successful and noteworthy *outcome* in order to qualify as a creative person.[2]

This broadened meaning conveniently enlists many Gatherer attributes such as planning and convergent and sequential thinking. In my opinion, this muddies the concept of creativity and *discriminates* against those who lack the means to bring their creative enterprise to fruition.

Let's say that one morning while brushing your teeth, you have an "A-HA!" moment that will revolutionize your company's product line. Your supervisor, a Gatherer type, who is too busy on his computer to listen, gives you a look of disbelief and continues on his not-so-merry way. Years later, someone else comes up with a similar idea, sees it through and wins a Nobel Prize. Does this mean you're not a creative thinker? I don't think so.

How do we define creativity? I believe it has to do with taking a mental concept, rotating it in one's mind and viewing it from different perspectives—like upside down, sideways and backwards! It's the ability to maintain a number of loose associations while still listening to the discussion of the main topic. This means the creative person may not get drawn as tightly into the ongoing conversation and may be accused of being unfocused or not fully participating (like the deficits found in so-called multitasking). However, the creative person is imagining other scenarios outside of the tight confines of the logical and sequential presentation to which he or she is supposed to be listening.

It's easy to confuse three distinct but similar-sounding concepts. *Multitasking* is an attempt to juggle a number of tasks rather than dealing with them one-by-one, in a sequential fashion. An example is talking on the phone, cooking and watching a television program simultaneously. Instead of cooking dinner, then calling Cousin Susie and then watching television, multitaskers move quickly from one to the other, all within the same time frame. We humans don't do this very well. *Continuous partial attention* occurs when we're only partially attending to lots of information and tasks at the same time. This heightened state of alertness to multiple levels of stimulation over several hours can create stress and mental fatigue. It's usually related to our electronic machines, because they're the only ones that can routinely hit us with auditory and visual distractions from every angle and every speed—simultaneously.

A third concept is *simultaneous multi-conceptualizing*. This is what people do when they come up with creative ideas. They are focusing on a primary situation or problem, but letting some other concepts flow in to their heads to bounce off of the primary concern. It's like taking a couple of features from the primary idea and stretching them out to touch new or related ideas.

Simultaneous multi-conceptualizing appears to fall somewhere between *multitasking* and *continuous partial attention.* It's like stretching a rubber band.

Creative genius.

We hear a lot about "thinking outside the box." What does that mean exactly? Think about a tightly wound ball of rubber bands. You're sitting at a table with five other people, staring at this ball and wondering about the best way to change, promote or somehow enhance its value to the company for which you work. While your associates continue to discuss the problem, you take one strand of rubber and pull it away from the ball.

You walk to several places in the room, where there are also rubber balls sitting on desks and encircle some of them with your rubber band. Some rubber strands from the new balls rub off on your strand. It stretches, almost to the breaking point, but does not break or snap back and damage the original ball of rubber bands. Now that you are at some distance from the table, where your colleagues are still talking, you can't hear them as well and they look somewhat changed. You walk back to the main table and re-attach the rubber band, along with the additional strands that rubbed off, to the first ball. The original ball is now composed of parts of other balls and one or more strands of the other three balls become the central focus, while the original ball becomes a secondary player.

As with most measurable characteristics, creativity exists on a continuum. There are linear Gatherers who believe in hard and true facts and tolerate little deviance from the norm. They exhibit only a mild degree of creative thinking and behaving, but this can be improved through effort. In the middle of the continuum we find individuals with both Gatherer and Hunter traits who have original thoughts that are a slightly clever twist on something that already exists. These insights result from a spark of creativity, along with logical associations and sequential thinking. Many people have the ability to be creative at this level.

On the extreme end of the creative continuum are persons like theoretical physicist Albert Einstein, mathematician and economist John Forbes Nash, Jr. and artist Salvador Dali, whose original, visionary and intuitive thinking seems truly natural and effortless.

They had the ability to come up with atypical, fresh ideas that offer unique perspectives. The initial reaction of others when first hearing this kind of idea is usually, "Where did that come from?" The second reaction is, "That's nutty." The third reaction is, "Why didn't I think of that?" A closer look at these creative ideas usually shows some logical connection to existing material, but these ideas really shift the frame of reference and bring a whole new perspective to the table.

Why do some creative people have difficulty fulfilling mainstream expectations? And why do their business associates and loved ones believe they need special handling in order to fit in? If a person has all the accepted mainstream gatherer-hunter traits and characteristics and is creative in addition, what's the problem? Who wouldn't welcome that additional innovative thinking power?

The problem is that in some cases creative insight is not a splinter skill piled on top of orthodox thinking, but rather represents an unorthodox *personality* that puts the person outside the mainstream. What are these out-of-the-box characteristics and where do they come from? In the second chapter, "Halvin and Garrison," Hal, the extreme Hunter, exhibited a few characteristics of an unorthodox, creative personality.

How to.

How can we encourage creativity? We need to do the same thing that fiction writers do, by constantly asking the question, "What if?" And we've got to have enough self-confidence to ask seemingly dumb questions in public. If you are intimidated, don't just blurt out your question. Instead, preface it with something like this: "I'm just brainstorming here and this might be a little far out, but—I was wondering if . . ."

We have to question everything and we have to give ourselves time and space to think (and *not* think). We can't succumb to information technology when it pushes us to go faster and faster while accumulating greater stockpiles of disjointed facts and details. In other words, we can't Twitter our creativity away. According to Shunryu Suzuki in *Zen Mind, Beginner's Mind*, "If your mind is empty, it is always ready for anything; it is open to everything. In the beginner's mind there are many possibilities; in the expert's mind there are few."[3] The headline for the Rolex Mentor and Protégé Arts Initiative states: "Listen, and then ignore everything. There are no firm rules. If you find any, break them as soon as possible."[4]

How not to.

If creative thinking is a natural process, we must remove the obstacles to creative thinking. Control Business Advisors, Inc. states, as reported in *Creativity in Business* by Michael Ray and Rochelle Myers, "The small businesses' three worst enemies: thinking too big—thinking too small—thinking too much."[5]

In the article "Creativity Killers" by S. Dingfelder in the *American Psychological Association Monitor on Psychology*, "Harvard University's Teresa Amabile, Ph.D., reveals that some of the worst offenders include: Fragmented work schedules [and] frequent interruptions, whether by meetings, phone calls or e-mails.... 'Interrupted people are unable to get deeply involved in the problems they are trying to solve'" and "A focus on short-term goals" is a creativity killer. "'A long-term visionary outlook'" is necessary.[6]

Time pressure also hurts. Workers use their mental resources to finish tasks quickly and clear their desks, rather than generating new ideas. A rushed rat will not take time to explore the maze and I don't think playing computer games will give the answers needed.

Interruptions and too much focus on detail can prevent creative thinking and even obscure the big picture. I'm reminded of a television advertisement at Christmastime, which showed a couple waking up to a wonderful surprise, a beautiful new car wrapped in a red ribbon, sitting right next to the Christmas tree in their living room. The woman waves her arms in the air, overjoyed. She exclaims, "Honey, look at that! Where did you find that big ribbon?"

But you Hunter types out there should remember that without knowledge, which is fact-based, along with careful planning and procedures, you'd be in big trouble. You need those huge linemen up front to protect you long enough to throw that touchdown pass. To be successful, whether it's a family, business or brainstorming team, it's important to have a balance between Gatherers and Hunters. A recent spate of books has told us to rely on our gut-level intuition. This notion is appealing because it upends our traditional belief that reason is more important than "vibes" when it comes to complicated decision making.

Blinkers and thinkers.

But these books do not tell us which people are more prone to blinking (Hunters) and which are more prone to thinking (Gatherers). One example given in Malcolm Gladwell's insightful book *Blink: The Power of Thinking Without Thinking* is how Confederate General Robert E. Lee outfought General Joe Hooker even though Lee knew far less about Hooker's army than Hooker knew about Lee's troops. "Lee won the battle because he knew less than Hooker." This example of a natural research paradigm supports the contention that sometimes "less is more."[7]

Another example of a winning general who trusted his instincts occurred in 1950, when General Douglas MacArthur commanded American forces in Korea. According to author David Halberstam, in his book *The Coldest Winter: America and the Korean War*, MacArthur decided on an amphibious landing at Inchon that would encircle the North Korean army and cut off its escape. The Joint Chiefs in Washington, persons MacArthur referred to as "small-bore bureaucrats," warned the general not to take that terrible risk. In fact, MacArthur spoke, himself, of Inchon as "a five thousand to one shot." Then he relied on his instincts and ordered the landings. It was a phenomenal success. Halberstam wrote that two weeks later, despite good intelligence and all facts pointing to massive intervention by the Chinese army, MacArthur's instincts told him to advance—what his generals referred to as "an arrogant, blind march to disaster." As a result of blinking without thinking, 33,000 Americans died and 429,000 were wounded before the war ended. Maybe the question shouldn't be whether to blink or think, but when to blink and when to think.[8]

Like most Gatherer scientists, psychologist B.F. Skinner, the father of behaviorism, had little sympathy for the creative personality. He believed creativity itself was the result of random "mutations." He compared creating a work of art to conceiving and birthing a child. According to author A.J. Toynbee in his book *On the Future of Art*, Skinner argued that in both cases the creator is no more than a locus through which environmental variables act and that the creator "adds nothing to the creation."[9]

IT creativity?

Authors John Palfrey and Urs Gasser champion computers and the Internet as promoters of creativity because they permit youngsters to mix and match and edit material from Web sites to create new forms of expression. Because of highly localized production, creative writing and performing are close at hand. Photos, videos and music are easily amenable to modification, remix and circulation. In this way, youngsters can use some degree of creativity and pull disparate information together, produce it and send it out to others. Palfrey believes this satisfies the more stringent definition of creativity and is certainly superior to passively watching television.[10]

Palfrey uses William Shakespeare as an example of a writer who used other people's ideas and information to help him create his plays. Granted, it's unlikely that any human has ever created anything out of nothing. Some writing teachers claim there are only thirty-six plots in all fiction. Writers "merely" find different ways to "dress up" these universal plots. I believe there's a wide gap between Palfrey's concept of mixing and matching and true creativity. If Shakespeare were merely mixing and matching, we'd have had a few hundred Shakespeares, not just one. Some critics have referred to profile creation on MySpace as "copy and paste" creativity.

One person who wouldn't be real happy with Palfrey's idea of creativity is a genius who has a museum named for him close to the street where I live. As stated in *The Persistence of Memory: A Biography of Dali* by Meredith Etherington-Smith, "If intelligence does not exist at birth, it will not exist at all. If men go on dying, blame Jules Verne, he was logical. To be logical is to be cuckolded every day by truth and ugliness. What is truth? Less than nothing. What is ugliness? Ugliness is that which cuckolds you every time with beauty. What is beauty? No one knows yet, since it is too obvious. Instead of writing a history of art, I am writing the art of history, since all our art historians are average cretins with the exception of your humble servant," said artist Salvador Dali, who liked to watch clocks melt.[11]

Mathematician and economist John Forbes Nash, Jr. is featured in the motion picture *A Beautiful Mind*. He talked about an epiphany with a flash of light and the fog moving away. His Hunter side came up with creative ideas and his Gatherer side supported them with evidence necessary for him to win the Nobel Prize, based on a cooperative bargaining theory. In his theory, there can be several winners in bargaining and not just one, as proposed by philosopher Adam Smith.

Hike!

The legendary football coach, Knute Rockne of Notre Dame, was sitting in a Chicago nightclub watching chorus girls dancing when his cell phone vibrated and he got an urgent text message from a student-athlete he was trying to recruit. Wait. Now that I think about it, 1930 was a little too early for cell phones and text messaging. So instead of having his head down *Twittering* a student-athlete or fellow diners with important questions such as, "How did you like the horseradish sauce on your bratwurst?", he happily concentrated on the dancing girls. As he lazily watched the show girls, it dawned on him that the dancers' movements might generalize to football. On the spot, Rockne conjured up what was to become the Notre Dame shift. What would college fans do today if we had no T-formation, no forward pass? Thank goodness, the Hunter personality's love of rhythm and music wasn't compromised by machines. No Digital Pandemic for the Rock, I'd say. Sometimes linear logic can lead us astray. The Gatherer director of the United States Census Bureau in the late 1800s stated, "Patents will soon be a thing of the past, as everything worth inventing has been invented." Why aren't Gatherers creative? Why don't they try new things and why do they seem negative about their serendipitous Hunter friends? It may not be in their genetic makeup to begin with, and focusing on details and facts doesn't lubricate the creative making machine.[12]

5

Lost in the Game World

At least 97 percent of twelve- to seventeen-year-old American adolescents play video games and one-third play mature games designed for adults. Each year, Americans purchase more than 220 million games, nearly two games for every U.S. household. In the United States, the video game business is larger than the motion picture industry. Americans spend more on video games than they do on movie tickets.

In Mizuko Ito's study, her group of anthropologists conducted an extensive survey of youth and information technology. They found that 24 percent of kids played on a daily basis and two-thirds had game consoles prior to the age of ten. Ito made it clear that this research is a starting point and little work has been accomplished to date on how different genres of games interact with gender, age and social class, but this does help give us general ideas of what's going on "out there."[1]

It's not just in the gaming area that we find a great explosion of products. A recent advertisement for a 3G cell phone bragged that the phone includes a link with Facebook to help us check status updates. One program doubles as a steering wheel in a high-speed racing game, while another allows us to stay atop of the world's financial markets, yet another application gives us access to all the music in our library and to one of the most addictive, logic games available.

Those are only five of the fifteen apps (applications) available on that phone. Other applications let us manage our banking, advise us on where to eat, identify the name of an artist or album, give us news from the various sources, let us make purchases, help us with foreign languages and entertain us with games that "test how you'd make it in the primordial ooze." Of course, programs to do "regular" tasks like texting, e-mail, calendar, contacts, digital music and phone calls are available as well.

It doesn't stop with businessmen and women, who may be able to handle all these options without getting a headache or plowing into a car on the freeway. Nine- and ten-year-olds can give you chapter and verse on their latest equipment—in a flash, especially the Gatherer types. One ten-year-old told me about his MP3 touch screen. "I've got app store games, e-mail, a calculator, a note pad, weather, clocks, address book and can go online if they've got Wi Fi. Battery only lasts an hour or two, but I've got a one-year warranty." He turned the device ninety degrees to the left and the screen adjusted so that the columns of numbers still faced him, vertically. I asked myself, *What does it all mean? Is it important? Can it teach us anything significant and will it really help make us a better people?*

According to Evan Ramstad's article "Coming to Tiny Screens All Over the Place," "Watching TV and movies on cell phones is so common in South Korea, people no longer think twice about it." Since 2005, South Koreans' use "portable devices that pick up TV broadcasts sent on a special frequency, a system…known as digital multimedia broadcasting, or DMB."[2] If just talking on your cell phone while driving greatly increases your risk of an accident and possibly death, I don't suppose watching movies like *Vanishing Point* and *Crash* while driving will do much for your safe driving record, either.

A full page insurance company advertisement asks: "Remember all the stupid things you did behind the wheel when you were a teenager?" Then it goes on to say, "Now add a cell phone, a vanilla soy latte and an MP3 player."

No one doubts that the IT explosion is here. The real question has to do with the effects of IT on self-concept, family, values and learning. A good number of writers believe that younger people, whom many call "Digital Natives" because they grew up with electronic systems and had early access to the Internet, are the key to our future. They believe that IT gives us a cold.

The Industrial Revolution led to sweatshops where seven- and eight-year-olds worked seventy-hour weeks and the absence of automobile seat belts led to huge numbers of unnecessary deaths. We'd like to skip those kinds of outcomes this time around, but the use of cell phones in automobiles has already shown that it may be too late.

Playing at learning.

James Paul Gee is a linguist and professor of Literary Studies at Arizona State University. His book *What Video Games Have to Teach Us About Learning and Literacy* takes a highly detailed, exhaustive look at video games. Gee finds electronic learning to be compelling and full of intrinsic rewards. He contrasts this new type of learning through gaming with traditional schooling, which he implies is mostly "skill—and drill." He believes some children can learn well in a traditional, uninspired environment but that's only because they trust authority figures, such as family and teachers, who have told them this drudgery will be beneficial in the long run. Then he states, "Other children have no such trust. Nor do I."[3] My experience indicates that children who distrust parents and teachers are usually headed for trouble. (I wonder if his favored electronic games will provide students with sufficient literacy to read his book. Then again, by the time our present six-year-olds reach college, they may not read books.)

A whole new generation.

This whole electronic revolution, with its emphasis on generational differences, is reminescent of the 1960s and 70s, but this time the goal isn't peace and free love as much as unfettered, self-directed pleasure (and learning?). Well, if you're a kid and you don't trust adults, it's likely you're headed for trouble, big time.

An over-focus on generational differences is fraught with exaggeration and stereotypic thinking, much like gender differences, and usually comes to no good. An example of one program is aimed at settling down rowdy teens. Adults can use their super cell phones to emit an annoying high frequency sound. It irritates teens, but older folks are spared the noxious bombardment because of their natural loss of high frequency hearing. Now the machines are *attacking* us!

Will cyber kids and young adults really come up with their own rules? David Brooks, syndicated columnist for *The New York Times* states, "In this way of living, to borrow an old phrase, we are not defined by what we ask of life. We are defined by what life asks of us. As we go through life, we travel through institutions—first family and school, then the institutions of a profession or craft… New generations don't invent institutional practices. These practices are passed down and evolve. So the institutionalist has a deep reverence for those who came before and built up the rules that he has temporarily taken delivery of."[4]

Gilbert Keith Chesterton, the great British philosopher, agreed. In his book *Orthodoxy: The Romance of Faith,* he explains, "Tradition may be defined as an extension of the franchise. Tradition means giving votes to the most obscure of all classes, our ancestors. It is the democracy of the dead. Tradition refuses to submit to the small and arrogant oligarchy of those who merely happen to be walking about. All democrats object to men being disqualified by the accident of birth: tradition objects to their being disqualified by the accident of death. Democracy tells us not to neglect a good man's opinion, even if he is our groom; tradition asks us not to neglect a good man's opinion, even if he is our father."[5]

Ito's research group also warns us to be "wary of the claims that there is a digital generation that overthrows culture and knowledge as we know it and its members' practices are radically different from older generations' new media engagements."[6]

New coach, new team.

This "pull yourself up by your own shoestrings" approach also ignores what I believe is the important role of mentors. A mentor is a person we admire who takes a special interest in us. As we grow up, this mentor (usually a parent, teacher or coach) engages our attention and draws us into the learning equation. To educate means to inspire and to "draw out," not to inundate the learner with externally rewarding entertainment scenarios that might, as a by-product, advance our thinking in some random way. What about our cold-blooded computer? Can IT systems replace the teacher, parent and coach? Not likely.

Some argue that electronic games are unique in satisfying the conditions for apprenticeship. Students can certainly spend an enormous amount of time at their computers, and they are working with materials designed by master computer whizzes. They are also given continuous feedback until they achieve a certain score or acceptable level before they go on to the next one. Much of this mimics the apprenticeship situation.

This analogy falls down in several areas, however. The motivation and intention of the apprenticeship master instructor is to teach important, specific skills. The designers of electronic games have not set out to teach specific skills, although some learning undoubtedly takes place. In the case of electronic games and even Internet visits to Web sites, the game player or student is choosing what, where and when to learn, not the apprenticeship master.

The apprenticeship is also a one-on-one, person-to-person relationship, and the feedback is not calculated to reward and stimulate as much as to prepare. Certainly the electronic game rewards the "apprentice," but do these nifty rewards have the staying power of human interaction, job application and job success? The master does not view the apprentice as a customer, but rather as a highly motivated student who will sacrifice all to learn the secrets of the trade. In the customer model, the customer is king and one never irritates the king.

More than it seems.

James Paul Gee recaps how good video games engage players with powerful forms of learning. How many video games are really good is another question, but let's take a look at a couple of Gee's twenty-nine learning principles. Number twelve is the Practice Principle. "Learners get lots and lots of practice in a context where the practice is not boring (i.e., in a virtual world that is compelling to learners on their own terms and where the learners experience ongoing success) They spend lots of time on task."[7]

Two thoughts come to mind here: First, as we will see later, some boredom may be a natural part of the kind of sustained learning that helps us to generalize. Second, just how much time is spent on task? Is the amount of time spent on these games really efficient in terms of learning these "principles"? And are these "principles" the best ones available for teaching and learning?

Gee's Intertextual Principle states that "The learner understands text as a family ('genre') of related texts and understands any one such text in relation to others in the family, but only after having achieved embodied understanding of some texts. Understanding a group of texts as a family ('genre') of texts is a large part of what helps the learner make sense of such texts."[8]

I think Gee is saying that after players have played different games and read the texts associated with them, they realize that these texts share some commonalities. Of course they would. But did any teaching principle intrinsic to the games help them learn to generalize from one text to another or did they already have this capability?

I believe it's the latter and recognize that one book is like another book; some of the strategies carry over to another text and another game is probably learned through comparison of the accidental (obvious and superficial) characteristics the two books share. This is learned in early childhood. Learning to tie knots in the Boy Scouts is an example. Recognizing these similarities a few more times is *unlikely* to strengthen the ability to generalize.

I think Gee is putting every thought and action under a microscope to justify gaming as a sophisticated learning model. Most of these thoughts and actions can be found in almost any activity in which we engage. Starting with electronic games and then trying to find possible learning outcomes is putting the horse before the cart. Why not start with the best practices in learning, which have accumulated over the years through experience and outcome studies, and then compare them to what is learned through electronic games?

For example, instead of starting with a balanced diet for our soldiers, why don't we just start with potato chips. As far as I know, these delicious salty snacks weren't designed for nutritional purposes. I've yet to see any in my local health food store. They were designed as irresistible and somewhat addictive snacks to make people happy and earn big bucks. But if you had an army in the field that was deprived of sodium, these crunchy little chips could satisfy that need.

Although we're not going to review all of Gee's learning principles, nor all of the fifty or more "Best Practices in Teaching," let's focus on some major points. "Best Practices" warns against too much reward and praise because teacher praise or the expression of judgment tends to foster approval-seeking rather than independence. Professor at North Seattle Community College Tom Drummond states in *A Brief Summary of the Best Practices in Teaching* that Halting Time is one technique that gives learners time to think or carry out directions when dealing with complex material. Guided Lectures, Storytelling, Modeling, Role-Playing, Peer-Teaching, Self-Talk, Nonverbal Expressions, Brainstorming and Fostering Learner Responsibility are just a few of these learning practices.[9]

It could be argued that none of these highly valued learning and teaching principles are found in electronic game play. Drummond paraphrases ideas from philosopher John Dewey's *Democracy and Education: An Introduction to the Philosophy of Education*: "The teacher, as leader, brings a mature view of learner development, *which will hopefully unfold over time*, and brings a thoughtful perspective on the long-term aims of this educational endeavor. The teacher has experience in the evolution of knowledge, skills and dispositions that *lay beyond the learner's awareness.* The teacher also brings his or her evolving understanding of the relation of the current study to *what it means to be human.*"[10] (My italics)

On the road again.

For example, follow the imaginary trail of three people who are lost in the woods. We'll see if they're learning anything. Look—here's where they stopped and walked in circles. They must have been deciding whether to try and cross the stream at this point, where it's about five feet deep, or to walk fifty yards downstream to some rocks. Stones have been brushed aside, so they threw pebbles into the water to determine its depth. Some leaves were pulled from a bush. They must have placed leaves into the stream to gauge the strength of the current and to help predict where they might end up if they lost their footing and were carried away.

We follow their footprints down to the rocks that form a natural bridge over the stream and notice they picked some wild berries. They had to decide whether the berries would make them sick and weigh that against the problem of starvation. A few partially bitten berries are on the ground and most of the other picked berries were discarded. The berries must have made them sick.

Think of all the learning principles that could result from those few minutes and that brief walk. The lost campers had to feel the weight of the pebbles and choose the correct color to ensure they could be seen at the bottom of the stream. The campers also used visual acuity and depth perception to size up the stream's depth, and they had to consider the laws of physics and geothermal variables to determine the direction and speed of the current. They needed to decide whether to go to the rocks and lose valuable time or cross quickly and possibly drown. The setting sun and a chill breeze from the northeast gave them information on when it would get dark and their exposure to the weather. They had to weigh bitter fruit against starvation.

To do these simple things, during their fifteen minute walk they had to invoke teamwork, leadership, planning, psychological factors and assessment of both long-range and short-range goals. They used verbal, auditory, visual-motor, gross-motor, balance, kinesthetic skills and all five senses.

As with electronic games, we could extrapolate and construct principles of learning from our walk in the park, but would they represent a powerful new approach to learning? I don't think so. Analogous to video games, the goals and best principles of learning weren't established before the hikers got lost. We don't know if the solutions they took were the best ones, because they were never tested. Isn't it a backward approach to establish the things we learned *after* the fact and then advance them as a powerful learning paradigm? At least this lost persons exercise was a lot quicker than electronic games...and they were able to stop and smell a few of the roses along the way.

Gee indicates that video games offer players strong identities. Players learn to view the virtual world through the eyes and values of a distinctive identity. One has to question whether this identification with a made-up character will strengthen one's self-concept and carry into real-world experiences. The process of assuming another identity during a game sounds like role-playing, which has been around a long time and has the clear advantages of interaction with real people in the real world. In fact, that is the basic underlying weakness of the video game learning paradigm: it's a game; it isn't for real.

Gee believes games make players think like scientists. I assume he means "encourage." Video games appeal more to the Gatherer than to the Hunter. That's because the Gatherer is mechanically and sequentially inclined. Basic games teach these pursuits at the expense of Hunter, intuitive strategies. Hunters get excited about violent games, supposedly designed for adults. As yet, there is no evidence that these learning paradigms generalize to the real world in a helpful way,and some studies indicate that academic achievement is negatively related to the overall amount of time playing video games. [11]

These games, according to Gee, let players be producers and not just consumers. In some open-ended games, players co-design the game through unique actions and decisions. I think we are all for that, depending on how long it takes and if it's the most desirable approach, but this is more applicable to older kids and the main concern here is the formative years. For younger children, how do these "producer" games compare to playing cops and robbers outside in the fresh air? Perhaps studies comparing one form of entertainment, video games, with other forms of entertainment such as checkers and chess would be beneficial.

As kids, many of us came up with hundreds of scenarios during outdoor games and even produced plays. The more complex games enjoyed by older teens and young adults may be useful in some narrow vocational areas or even in basic problem-solving skills, but considering the low density of electronic games (amount of time needed to learn a concept), wouldn't youngsters learn faster and better with well-designed learning paradigms?

Flying low.

Flight simulators are an almost exact reproduction of the airplane cockpit. While these training devices can be viewed as very expensive electronic games, costing millions of dollars, and as they have been around for at least fifty years, no one has claimed that hundreds of hours in a simulator, or even hundreds of hours flying an airplane, generalizes to scientific thinking or broad, real-world skills.

After all, despite thousands of hours flying, pilots don't get Ph.D.s with their wings. The major difference between simulators and electronic games is that with simulators, a very specific need is identified first, i.e., to fly a certain airplane. Then, the electronic game or simulator is designed specifically to strengthen one's skills for that narrow yet complex task. I've been in advanced simulators and they can leave you in a breathless sweat. However, unlike other games, comes direction from experienced instructors by way of the simulator, not from the student pilot (which is precisely why it works).

We don't allow untrained people to pop into a multimillion-dollar simulator at any age, between five and ninety-five, and just start playing around. It could be argued that one would eventually learn to fly a simulator that way, but it would take an enormous amount of time (low density) and too many crashes. And there's no way to guarantee that the student's self-initiated program to fly the simulator is going to be the most effective one in the real world. It's unlikely that these "students" could take a real airplane into the wild blue yonder and get back in one piece.

There's no doubt that intelligent and experienced designers have programmed electronic games and that they contain rules and directions that if followed will lead to high scores. But the difference is this: The simulator has been programmed to teach specific skills and emergency procedures based on the experiences of thousands of experienced and successful adult pilots. The toy gurus and game manufacturers have a different goal. They want their customers to be entertained, feel important and successful, and go on to buy more and more games. The bottom-line goal here is money, not learning.

Fail-safe.

Another advantage to video learning that Gee brings up is the fewer consequences of failure. With IT, players are encouraged to take risks, to explore and to try new things. Young children are given the opportunity to determine, independently, when the game is getting too scary and pull back to a safer level. When players fail, they can start over from their last saved game. Okay, he's right. Some children may pull back, but I have seen lots of kids, in many areas of life, who have disregarded danger until they were in over their heads, resulting in heightened anxiety, fearfulness and post-traumatic stress.

This paradigm is reminescent of the approach-avoidance paradigm we used in a physiological psychology study to test stress in rats. The professor with whom we clinical psychologist types worked designed the study in which delicious morsels of food were placed at the end of a runway, but as the rat got closer to the reward, shocking mechanisms beneath the runway increased in intensity. The closer the rat got to the goodies, the more it wanted to avoid them. The brutal next step was to nail the rats to boards and put them in a refrigerator overnight. The next day we were told to cut the rats open and check their organs for wear and tear.

Some of us felt sorry for the rats and asked if we could tape them down, but the professor was adamant. "They're only animals," he said. Sure enough, the stress was physically evident. If, as Gee states, the games are about social interaction and competition, wouldn't the child want to push himself or herself beyond their perceived borders of safety in order to not "chicken out" and in order to compete successfully?

This seems like an approach-avoidance conflict to me. And nobody's there to supervise and protect little Heather. She's on her own, responding to the enticements of ambitious creative geniuses and designers who are motivated to speed her down the runway as quickly as possible so they can get *their* reward: money.

Cops and robbers.
This real-world game also encourages risk-taking and provides a way out. "Bang, bang, I got you!" Or, "I zapped you with my laser gun." "No, you missed. I'm only wounded and can shoot with my other hand," etc. The same is true in table games where one risks greater or lesser purchases of land and hotels or in such card games as poker, except in these games the player is learning in the real world not a virtual world. The player is seeing, feeling, sensing, smelling and perceiving subtle changes in facial expression and body language that can't be reproduced in a computer game.

Teacher instruction is faster and better. Good teachers keep their students at the forward edge of their accumulated learning curve. If the work is too easy, learners may have fun—until boredom sets in—but they're not learning much. If the work is over their heads, they will become inattentive and lose valuable time. Pace, or movement over time, is a key word that comes to mind. Everyone could become a clinical psychologist if they had twenty years to master the requirements, not the ten to twelve years expected. But that would be enormously expensive, among other problems, and just wouldn't work.

The real question is this: Is time better spent, especially for preteen children, sitting for hours alone in front of a screen learning some of the things we learn in school and from such other games, but learning them alone, in a virtual, unreal environment? There is no evidence, despite feelings of distrust toward authority and most teaching practices, that machines can provide instruction that precisely fits our students' personalities, interests and aptitudes.

Most kids learn well in an environment of human sharing and feedback, where good teachers individualize their instruction. Advanced learning classes and programs for the learning-disabled are also available. I doubt that all video games are effective learning tools and I know all teachers are not good ones. The solution is training or replacing the unhelpful teachers, not substituting them with electronic games.

Once upon a time, physicians prescribed massive doses of vitamin C following open heart surgery. Naturally, this created a great clamor and the general population started gulping down fistfuls of vitamin C. Unfortunately, later studies showed that high doses of this vitamin, unless there was an extraordinary deficit imposed by surgery or some other traumatic condition, did little good and resulted in some nasty side effects.

Computer learning—which borrows quite a few tricks from electronic games' methods of format, with its seductively motivating features—may aid some children with learning disabilities to cross thresholds that are keeping them from reading, spelling or doing math, but may not be the ticket for the majority of children who prosper in a warm, sharing, group atmosphere.

An article was published in the *Journal of Educational Psychology* where researchers Maria De Jong and Adriana G. Bus performed a study of emergent readers where adults read to twelve children from a regular paper book while twenty-four children explored an electronic book with similar illustrations and story content. They found that the regular book format was more supportive of learning about story content and phrasing, while both formats supported internalization of features.[12]

Monkey see, monkey do.

Lead specialist of Division and Research for the Virginia Department of Education Dr. Lawrence McCluskey helps us keep the human learner, not the machine, in the forefront of our research. He stated, "Technology simply offers a means by which one may process knowledge. It offers enormous potential for benefiting mankind, but people who believe that putting a computer on every student's desk will heal all educational ills would do well to rethink that proposition. A monkey at a computer terminal has roughly the same chance of writing *Hamlet* as that same monkey at a typewriter."[13]

If the proper software is designed and the adult instructor can enter a program that approximates the child's level of knowledge, pace, and best avenues of learning—visual, auditory, etc.—while offering warmth, encouragement and advice, computer learning can be a useful adjunct to help a child surge forward. The questions are these: Can electronic game design generalize to teaching in the classroom? And what will it cost?

What about the rate of learning? Will the child's hyped-up motivation decrease once the novelty wears off? If a circus dancer can learn to be a ballerina, all is well and good. But as with multi-vitamins, children shouldn't be alone in their rooms, gulping down electronic games. They're much better off interacting in the real world and learning social skills that are as important as, if not more important than, the three Rs they learn in school. And, as always, we must ask what qualities of human life may be sacrificed.

One additional caution for those of us who are interested in how technology can help children is that writers and researchers are coming onto this process as adults with mature, adult brains and the benefit of a broad education under the tutelage of generous and inspired role models and mentors.

We must remember that this is quite different from what's going through the mind of a five-or six-year-old, whose brain is not fully formed and who does not have the experience and wisdom to understand the seductive powers of video games and how they are altering his or her mind. I am again referring to my own "Developmental Principle" from the book *Parent, Child and Community: A Guide for the Middle Class Urban Family* by authors Herbert Goldstein, Patricia J. Pearson and myself, which points out that children are not tiny-sized adults. We need to remind ourselves of that again and again.[14]

The adult Gatherer may like the mechanical aspects of technology and the adult Hunter type enjoys the violence and adventure of first-person-shooter games. Wouldn't it be ironic if older, so-called digital immigrants (people over forty) eventually make up a large share of the video game market? The Gatherer type, who is mechanically inclined and who finds emotional release in fantasy movies or even science fiction comic books may fall in love with this virtual world in a heartbeat.

Back in the 1970s, this type of person would find his or her way to California for Rolfing, Encounter groups, Regression to the Womb therapy and other experiential venues that promised to release their repressed feelings. Perhaps today's gory action games will help some individuals let off steam.

Gee states that people of every ethnic group and social class enjoy video games and more girls and women are playing every day.[15] Mizuko Ito's carefully conducted—and more recent—survey supports this observation, but indicates that boys are more interested in the aggressive and violent games. Young males dominate recreational gaming. Poor kids are left out of the loop because of the cost of electronic games as well as other electronic devices. Those who are better off will always have the latest games and equipment, as is true in many other areas, and the poor children may go without. In this case, it could be a good thing.

Rollercoaster.

Some praise the active features of video games compared to the passive content of traditional learning. Ironically, research indicates that it is the active feature of video games that leads to more aggressive tendencies. Researchers studied the effects of a violent virtual reality game. Older adolescents were compared *after* they interacted with this virtual reality game, *after* they observed others who interacted with this scheme or *after* they simulated game movements, except for the violent ones.

Researchers have found that adolescents who *interacted* with the game were more aroused and had more aggressive thoughts than those who observed others play the game or those who simulated the game movements.[16] Mizuko Ito's group seems to agree that interactive aspects make recreational gaming more experiential than, say, watching television. A twenty-one-year-old talked about some of his favorite games. Describing them as pretty grotesque, he said the pink demons scared him. He used words like shocking, intimidating, frightening and said that they promote irrational fears and panic.[17]

Maybe violent video games aren't so harmless after all. Here we are, worried about old-fashioned television programs that massage us into passivity, extra-salty potato chips and all. But maybe the video game computer is like the Trojan Horse. It enters our home innocently enough and then sucks our kids into its seductive agenda. Our television set doesn't reach out, grab us and draw us to its inner core. Don't worry parents, little warrior armies don't come sneaking out of your kids' computers at night to run across their beds, but their interactive battles, with hordes of dreadful demons, could resurface in their dreams.

What concerns me is the idea of the five- or six-year-old entering and exploring a game world that perhaps does not have his or her best interests at heart. There is no way parents can know exactly where the game is taking the child at any given time. How many parents would turn their kids over to an arcade on the other side of Chicago, Los Angeles or New York City? Not many, I suspect. We would want to be there to monitor what kinds of games they were playing and we would be wary about other people, both young and old, who might be hanging around, drawn to the excitement of the games or using them as bait for devious purposes.

One of the most ballyhooed new games, featured on a cell phone, isn't doing so well commercially. It's only sold a million copies and it cost fifty million dollars to make. What happened on the way to the marketplace? According to a *Wall Street Journal* article by Ben Charny and Yukari Iwatani Kane, "'Spore was so big, so innovative, and so open-ended that it was difficult to communicate exactly what it was about,' said John Taylor, an analyst at Arcadia Investment Corp."[18] How innovative is it? Like graduate school physics or seventh-grade algebra? No one knows. Taylor's description sounds like some of the experimental design courses I took in graduate school. I wanted to quit too, but I wasn't there to be entertained.

"'Players wanted more action from the start," says Ted Price, president of Insomniac Games. "'We intentionally changed up the way we created the (single-player story) and introduced more highs and lows to ensure the player was taken on a much more emotional rollercoaster ride," according to a *USA Today* article titled "A triple play awaits fans of sci-fi gaming."[19]

I went to a starter game at one Web site after interviewing some five- and six-year olds. The site is free, and when I played the initial game, I received hundreds of points for merely clicking on some vertical and horizontal lines to connect a theoretical electric grid. I was also encouraged to bring my friends along to the site and to work toward exciting rewards.

It didn't take me long to enter the computer programmer's mind and see how he was laying out small rewards and drawing me more and more into the game world. Parents: Children are playing with some highly creative and clever professionals. They're a lot more interesting than parents or the teachers at school and they may be deeper into children's mind than children's teachers would ever hope to be.

My primary concern with video games and other electronic devices is their effect on young children whose minds are still quite malleable. Many of the really scary games are designed for older teens and young adults and are based on consumer demand, but our society has always recognized the vulnerability of children to physical and emotional abuse; that's why we have accredited schools and certified teachers.

Of course, some people see any imposition of structure as tantamount to censorship and brainwashing. I think conservatives and independents sometimes forget about the Developmental Principle: "Kids aren't midget adults."

There are parental guides for the *content* of video games, but not for the intellectual *process* involved in these games. Years ago, neuroscientists knew the brain matured gradually until the late teens, but once mature, believed the brain was fixed and not open to change. In other words, we were stuck with the brain circuitry with which we were born. In their book, *iBrain: Surviving the Technological Alteration of the Modern Mind*, neuroscientists Gary Small and Gigi Vorgan, also at UCLA, wanted to know if digital immigrants (people not ever exposed to the Internet) could train their brains relatively quickly. Using our friendly MRI, the researchers discovered that after just five days of practice, the same neural circuitry in the frontal area of the brain that was used by computer-savvy subjects was activated in the Internet-naïve subjects. "Five hours on the Internet, and the naïve subjects had already rewired their brains." The digital immigrants spent only one hour per day over the course of five days.[20]

This prompts researchers to ask: If our brains are so sensitive to just an hour a day of computer exposure, what happens to young people whose brains are much more malleable and plastic? What happens to their brains when they spend not one hour a day but up to eight hours each day, month after month, with their electronic games, smart phones and other devices? No one knows the answer to this question, but we know enough to suspect there may be some serious problems.

As a result of this work and other studies, Small and Vorgan fear that neural circuits that control traditional learning may diminish, and that over-involvement with and in game playing may lead to social isolation, diminishing the spontaneity of interpersonal relationships and reducing social skills.[21]

In his book, *Mozart's Brain and the Fighter Pilot: Unleashing Your Brain's Potential*, Richard Restak, Professor of Neurology at George Washington University Medical Center, describes a study in which monkeys were trained to use a specific finger for a particular task. When the animals repeated that task enough times, the number of cells in the brain representing that finger expanded at the expense of the inactive cells. The same area measured in the monkey experiments was also enlarged in blind readers of Braille. Activity in the area of the brain receiving impulses from the fingertips of the Braille reader varied over short time spans and when the Braille reader took a weekend off from reading, the corresponding area diminished. When the Braille reader returned to reading, that area enlarged once again.[22]

Professor of Cognitive Neuroscience at the Stockholm Brain Institute, Torkel Klingberg, reports on several pertinent studies in his book *The Overflowing Brain: Information Overload and the Limits of Working Memory*. One such study was done by Ryuta Kawashima, a Japanese neuroscientist, who in an unpublished study, "measured the blood flow in the brains of children in three different situations: while playing computer games, while resting, and while doing repetitive arithmetical exercises (adding single-digit numbers)." Kawashima found that the games activated primarily the visual and motor areas of the brain (seeing and touching), whereas the simple arithmetical exercises activated the all-important frontal lobes. During the video games, the children may have relied on simple short-term memory, which is just retention of repetition or information, whereas the arithmetic problems used working memory that requires a greater demand and activates the executive functions (orchestra leader) of the frontal lobes of the brain.[23]

Short-term memory is instant retention of something we hear or see. Working memory requires us to *do* something, such as retain and *use* what we see or hear. For example, if we order a steak medium rare, potatoes and a special glass of house wine at a restaurant, the waitress has to remember these items until she writes them down or perhaps steps back to the kitchen where she relays this order to the chef. But the question is will her working memory include the particular wine we ordered?

Some individuals think this so-called working memory is the basis for general intelligence, because it allows one to consider different angles of a problem simultaneously. Maybe one needs a good working memory before going on to higher levels of learning and thinking. This could be, but what about kids with ADHD (Attention Deficit Hyperactivity Disorder)? They are easily distracted and yet many show high intelligence and high achievement in life. [24]

A softball player on first base with one out and a left-handed pitcher throwing to a long-ball hitter has to hold and rehearse many visual scenarios. The first base coach gives her a number of options, such as taking a short lead off of first, sitting tight, going on a hit-and-run, stealing second and worrying about a double play, all while keeping her eye on signals from the third base coach. In other words, our working memory is not just storing information passively; it's *working* the information.

Kawashima also felt it was possible that these differences in activity patterns were related to the game's heavy demand on stimulus-driven attention that rewarded speed of response, but required little working memory. Conscious attention willingly forces us to do something, such as read a chapter in a book. Stimulus-driven attention is involuntary and pulls us in without our consent.

The pug in the park.

One day I was walking through a park when I noticed a cute little pug dog attacking something on the ground. As I got closer, I noticed that the pug was trying to catch a bit of reflective sunlight that flashed from the wristband of its owner and zigzagged across the sidewalk. Time after time the pug persisted in trying to catch the elusive light. This dog could not hold back or inhibit his responses (disinhibition). He could not generalize sufficiently to understand what was going on, place the target in a context or forget his earlier lack of success. He was basically stuck. This is an example of stimulus-driven attention.

In psychology we might say our cute little pug is "stimulus bound." The stimulus, or flashing light in this case, binds him up and he can't escape its hold over him. His frontal lobes, or lack thereof, don't allow him to inhibit. He is both *over-focused* on the light and *under-focused* on the world around him. Children with attention deficit disorder, whose brain chemistry diminishes the effectiveness of their frontal lobes, are not only highly distractible, but at other times, especially in one-on-one situations with an engaging stimulus, also can become *over-focused.*

This might explain the popularity of the term "electronic cocaine" when describing some children's stimulus-bound addictive response to electronic games. I don't believe electronic games lead to attention deficit hyperactivity disorder, and kids who suffer from ADHD can benefit from computer learning—as opposed to electronic gaming—up to a point. The strong visual feedback and earphones let them exclude extraneous noises and movement.

However, overuse of basic electronic games by kids who do not have ADHD can lead to inattention (not ADHD) and boredom when conversing with *real* people, working on "Best Practice" curriculum materials and even when viewing movies and television. Another example of being stimulus-bound is an adult patient who suffered from a frontal lobe injury. He stood in a cafeteria line piling food on his plate. Even after the food covered his pants and shoes, he continued to serve himself more helpings. He was stuck in a stimulus-bound mode.

Torkel Klingberg reported on some other studies where children seemed to acquire skills while playing basic electronic games. In the study "Improving Children's Mental Rotation Accuracy with Computer Game Playing," published in the *Journal of Genetic Psychology,* subjects who played a puzzle game for eleven days were better able than the control group to put the puzzle parts together. [25]

In another study, according to the article "Effects of Reducing Children's Television and Video Game Use on Aggressive Behavior," published in *Archives of Pediatric & Adolescent Medicine,* frequent players of computer games were found to be superior in several tasks measuring visual perception; they did better than the control group on identifying how many items flashed on a screen. They did it better and faster than controls, and this may have increased their stimulus-driven attention. This was in comparison to a control group that played video games rarely or never.[26]

Klingberg points out that just because basic games don't engage the frontal lobes doesn't necessarily mean they're bad for kids. He reminds us that lots of activities, including some sports, don't engage working memory.

I think these basic games don't seem to engage the frontal lobes, working memory or controlled attention, and the functions they seem to be teaching are relatively unimportant for abstract learning and even most vocational positions. If one works on an assembly line, sorting and piecing parts together rapidly, this is good training. It's almost as if the game people have accidentally stumbled into a simulator or learning machine process that teaches a few specific, if not complex, functions. It also reminds me of the original Gatherer, 8,000 years ago, sorting seeds while the Hunter prowled the forest for game. Are we going backwards?

This lack of stimulation of the frontal lobes is of concern. If young kids spend too much time with these games, they won't be exercising more important functions mediated by the frontal lobes. Our goal should be abstract knowledge, learning to learn and thinking skills. These basic games just don't cut it. The frontal lobes control attention and working memory and are vital in many team sports. Compare chess or board games with puzzle games. No comparison. If these basic games are restricted to just an hour a day for older kids and/or used as a reward for learning something substantial—such as reading comprehension—that is okay.

We need lots more research, including longitudinal studies where the same subjects are followed for years to determine true outcomes. We also need replication of current studies by other researchers in non-affiliated institutions. The game industry is rich and powerful. We need to be wary of researcher entanglement with companies that produce the games.

Small and Vorgan infer that the IT environment has resulted in what they call "a state of *continuous partial attention.*" (My italics) CPA is when we *stay* continuously busy while trying to keep tabs on everything, without ever really focusing on anything. This is not the same as multitasking, the authors point out, where we have a separate purpose for each task. They question whether staying in contact with friends on the Internet in the midst of experiencing this state of *continuous partial attention* leads to an artificial sense of intimacy compared to when we focus on one person at a time. When their subjects worked on the Internet for several hours without a break, the researchers reported frequent errors and noticed feelings of being spaced out, fatigued, irritable and distracted.[27] This concept has applications not only in the areas of learning and stress, but also in controlling behavior. Some people have such a high need for control that they can't prioritize and pull back from engagement with everything and everyone around them.

Related to the issue of CPA are studies of multitasking and dual tasking. We can drive our fuel efficient machine merrily along the highway, whistling a happy tune, singing a song, listening to the radio or even talking to other passengers without much impairment, but the effect of cell phone conversations is comparable to driving with a blood-alcohol level above the legal limit. It is estimated that talking on cell phones while driving results in 2,600 deaths and 330,000 injuries each year in the United States, according to an article by David L. Strayer and William A. Johnston in *Psychological Science.*[28]

Let's see if we can compare and contrast some differences between face-to-face and "Best Practices" learning and electronic games (at right).

Category one shows that the Electronic Entertainment and Communications (EEC) column is tapping primarily into visual and motor responses, while the Face-to-Face and Best Practices (FTF/BP) column is more abstract and engages the all-important frontal lobes.

Second, games are sequential in presentation. In the Electronic Entertainment and Communications (EEC) column, one goes from A to B to C to achieve rewards and win. Even when the player has several alternatives to consider in the more complex games, the entire system is geared sequentially. When you earn a set amount of points, you progress to the next stage or level. In face-to-face

BASIC ELECTRONIC ENTERTAINMENT AND COMMUNICATIONS	FACE-TO-FACE AND BEST PRACTICES LEARNING
VISUAL AND MOTOR	FRONTAL LOBES EXECUTIVE FUNCTION
SEQUENTIAL–SHORT TERM MEMORY	WORKING MEMORY, LOGIC, INSIGHT, EMOTION AND BODY LANGUAGE
FASTER AND FASTER	VARIABLE TEMPO
IMMEDIATE REWARDS	OFTEN—DELAYED REWARDS
LIMITED TRUE SELF DISCLO-SURE AND ANONYMITY	SELF—DISCLOSURE AND OPE-NESS
CAPTIVATING GRAPHICS AND SOUND EFFECTS—STIMULUS BOUND—STIMULUS DRIVEN ATTENTION	WORKING MEMORY
SIMPLE MEMORY	WORKING MEMORY
MULTITASKING AND CONTIN-UOUS PARTIAL ATTENTION	PRIORITIZING

learning, sequential approaches are sometimes used in classrooms, but with the availability of other areas such as art, music and social interaction in the lunchroom and on the playground, there are plenty of opportunities for non-sequential responses.

Speed and immediate rewards are necessary to elicit stimulus-driven responses. In electronic entertainment, earning rewards comes quickly, precisely and logically. In human interaction, even in the classroom, interactions are at variable rates. Person-to-person learning sets a mixed tempo. People speak quickly, then slow down or stop altogether. Face-to-face social learning is epitomized by the old army saying: "Hurry up and wait!'

Rewards are usually delayed in the face-to-face (FTF/BP) condition. This delay actually helps learning because one can't rely on quick outside reinforcement and must develop an inner motivation to persist. It also aids in the development of frustration tolerance and the postponement of self-gratification.

Self-disclosure is important in maintaining a healthy self-concept. It is not easy in the face-to-face situation, and openness and trust are needed for it to blossom. There is an expressed vulnerability in true self-disclosure that helps one develop as a person. The anonymous disclosures to "friends" and strangers via computer is eerily factual and distant, nothing at all like the emotional give-and-take of the real world. It requires neither high levels of trust nor openness. Research shows that we don't reveal our innermost feelings online.

In no other industry can one find such captivating graphics and audio systems. Only the film industry comes close, but the spectator in a movie theater is not engaged and interacting directly with the screen. Basic game design and graphics lead to stimulus-driven attention, not the purposeful attention necessary to master challenging but necessary subjects. The EEC process is more intense than, say, television, because of the interaction between the gamer and the game, but the breadth/scope of television programming is much greater than electronic gaming.

At the same time, we know that electronic entertainment is more engaging than television because aggressive content brings out more aggressive thinking with gamers than with television viewers, especially those who already have aggressive tendencies.

The anonymity of the computer encourages the disclosure of information from the private self that is rarely shared, except with perhaps a handful of friends and soul mates. Some persons never disclose thoughts and feelings from their private selves in their lifetimes, but are happy to share information from their social and public selves. When individuals share too much from their private selves too quickly, it is ordinarily seen as a sign of emotional distress and is not usually welcomed. Others see it as intrusions into their personal space and as the lack of well-defined psychological boundaries.

The use of technology encourages multitasking and the evidence is becoming clear that the human brain was not designed for this. Perhaps if electronic games alter the brains of our five- to six-year-olds, we will have adults in the future who are highly adept at multitasking, but the losses suffered from the inability to focus and prioritize could be great.

Granted, because of the novelty and immediate gratification of IT, we may be able to learn and memorize new information, but we may be *unlearning* the very process of learning itself. I believe the true learning process requires concentration, patience and inner motivation. If we begin to expect constant novelty and immediate reward, what will happen if we carry these learning expectations into other areas of our lives?

If these expectations become ingrained, we are likely to experience boredom in the ordinary learning situation. We hear this expression a lot these days, especially among teenagers. In fact, boredom may be the *"default" mode* for teens. Notice how the machine is seeping into our lexicon? Boredom: what is it? We usually think of boredom as a lack of interest. But why would a student doze off while reading about the Aztec nation, its battles with invading Spaniards and daily human sacrifice?

The student may be expecting a hyped-up situation with incredible visual feedback and the "gratifying" opportunity of killing scores of Aztecs and/or Spaniards by simply moving a mouse.

Students often play an electronic game in the evening, sometimes late into the night. They report feeling a rush and believe their successful strategies give them important new ideas. But the next day, these electronically-induced learners have buyer's remorse when they realize they've forgotten much of what they learned—and forgot to do their school work. It didn't seem nearly as important or fulfilling as it had the night before.

Games Parents Should Play:

1. Gatherer parents: Teach your Hunter kids how their electronics work. Don't let them skip over fundamentals to get to the action. It will help them later.
2. Hunter parents: Teach your Gatherer kids to question everything about the games they play. How can they create and improvise? Game rules are made to be questioned.
3. Search for "frontal lobe" games that require working memory and thought. Don't let your child be stimulus-bound.
4. Parents: Learn how to play electronic games. If you enjoy them, play them with your children. If you don't enjoy them, play enough to know what your kids are doing and what they're learning or not learning.

6

The Vital Regimen

Family Cohesion

Dog slobber.

Now let's see how machines impact the lives of the Gatherer and the Hunter at home. Heather is a five-year-old Hunter. On a quiet Sunday afternoon she watches her three brothers play electronic football in the family room. Her oldest brother, a Gatherer, lies on a couch under a soft quilt, while the other two sit at opposite sides of the room. More than an hour goes by without sound or movement. Little Heather's not happy. Hunters are like that. (Fortunately, her healthy immune system still resists the pandemic incursion.) She misses her brothers' attention and is still innocent enough to react in a natural, emotional way. Hands on hips, she pulls herself to her full height of three and one-half feet and shouts out her disdain: "That game is just dog slobber."

This outburst garners no response from her siblings. Her oldest brother digs himself deeper into the couch. He feels as if he's floating, detached from his body. His only physical sensations come from his hands as they manage the game and the visual imagery of moving football players. He feels like a robotic man inside a machine *and he likes the feeling*. But while he's locked in there, he's not getting much exercise. Even now, his arteries could be as thick and stiff as those of a forty-five-year-old. The next oldest, a Hunter, gets up and skips through the house while still pressing buttons on his game pad.

Hands off.

Heather's only recourse for entertainment is her computer. She uses the stylus to draw two figures riding bicycles and seems pleased with her efforts, but realizes at a deeper level that something's missing. The explosion of primary colors fills her with awe and she feels momentarily important, but underneath she's distressed. With her old coloring books she could feel the pressure she exerted on the crayon and run her fingers over the wax markings on the paper. She liked the smell of the crayons and the texture of the paper. The drag of the paper anchored her movements from left-to-right, right-to-left and in circles. When she finished drawing and coloring she could use her scissors to cut out the pictures. Then she'd use tape and tacks to proudly hang her artwork on the kitchen wall for her mother to see.

Too bad her mother, a Gatherer, threw out the coloring books when the new computer arrived, because they were "no longer necessary" and messed up the family room. At age five, Heather already knows it's a plastic world. Fortunately, she's had a few hundred-thousand interactions with humans before touching her first machine. She's still innocent, but the machine is ready and waiting to tempt her. A modern-day version of the serpent in Eden, it offers Heather the chocolate-coated fruit of unearned pride.

Heather and her father, Hal, are Hunters. Hal stops doing some work he's brought home from the office, walks into the family room and flips on the television. Heather comes in and joins him watching a baseball game while playing catch with an old tennis ball. Heather's brother Garth, nine years old, is a Gatherer. He stopped playing electronic football and is watching the game, but he's viewing it in an adjacent room, on a spare computer. Garth prefers the computer over the television because of real-time statistics.

Data is updated between each pitch to tell Garth the probability of a hit and whether the ball will go to left, center or right field. It documents the type of pitch and how the pitcher does against the left-handed batter who's up right now. The pitcher's earned-run average and the hitter's batting stats are also calculated after each pitch. This way he doesn't have to learn to calculate these baseball stats on his own. That would require effort (and thinking).

Besides, it's a lot "cooler" than sitting with his dad and sister. Gary's losing valuable interaction time with his father and sister, but he's happily in the clutches of computer land and growing less and less dependent on real-time people. When Hal goes to the kitchen to get snacks, he hears a bloodcurdling shriek and returns to find Heather clutching a bruised right shoulder. Apparently she got too close to Garth's computer and he reacted physically.

Hal growls his displeasure and sends Garth to his room. *To his room?* Even before reaching there, Garth has his hand on his cell phone. It's packed with an Internet browser, e-mail, video player and MP3 player. Gary is off and running.

Let's discuss what's happened here. A nine-year-old boy is disciplined for physically abusing his sister and sent to his room to think about what he's done. Two things result: First, immediately after committing a negative act, Garth is immersed in sheer pleasure, and in this way, he is actually rewarded for a negative act. Behavioral psychologists have demonstrated time and again that rewarded behaviors will increase in frequency. Garth is being taught to clench his sister's shoulder when she is a pest and he also learns that impulsive negative responses are rewarded.

Second, Garth has no time to absorb emotions such as guilt, to think about his behavior and how to correct it, or how to better please his father, whom he truly loves. This is worse than suppressing his thoughts and emotions. Garth doesn't say to himself, *It was her fault and I don't agree with my dad's concerns and discipline.* No, Garth doesn't even confront what happened. Because of emersion in machines, feelings get buried in a millisecond, at subconscious levels. These feelings may not be accessible when Gatherer Garth needs them, and they may someday result in behaviors not under his conscious control.

In addition, Garth learns to turn to a machine when frustrated. No development of frustration tolerance for Garth. He doesn't have the time, and no child has the inclination. If Hal had taken the time to discuss Garth's actions and examine the sequence of behaviors that led to his negative behavior, it would have benefited Garth greatly. Being a Hunter, it's unlikely Hal would do that. He's into the big picture and makes snap judgments. At least he could have relieved Garth of his cell phone and made him sit in a chair in the hallway for fifteen minutes. Then Garth could have stewed on things for a while.

And what about family cohesion? When one ten-year-old boy in one of my schools some time ago was acting out at home and at school, I advised his dad to choose a short period of time during the day or evening and give his son his undivided attention. It didn't have to be fancy. They could dig holes in the backyard or just hang out, for all I cared. "Keep it simple," I said. Turns out, his dad took him to a football game where he proceeded to listen to the game with a portable radio stuck in his ear and watch the action on the field with binoculars. His son was left out—and this was long before the Internet and IT.

Today, Mom and Dad can peer at their machines in their separate home offices—or anywhere else in the house—while the kids peer at their machines in their separate rooms—or anywhere in the house. Remember, the family that lives together stays together, while the family that lives apart, stays apart. Without embedded face-to-face relationships, it's not a family; it's a collection of individuals.

Self-concept and "friends."

Garrison Graham, a thirteen-year-old Gatherer, is a bit confused. No, he's more than confused—he's downright anxious. He knows who he is when he's on the family couch with earphones stuck in his ears or slumped over his computer, but he's not so comfortable with friends and peers. That pretty blond freshman, that nasty teacher, people who don't like him…for no apparent reason. What does it all mean? He feels unprepared and splintered into a thousand pieces. He'd give anything to pull it all together, to find himself and hang on to a single self-image for more than a few minutes at a time.

There are some people who think Garrison's in great shape such as John Palfrey and Urs Gasser, authors of *Born Digital.* They view the digital society as an exciting time when young people can use social networks to "learn what it means to be friends, to develop identities, to experiment with status and to interpret social cues."[1] Are we talking out of touch here?

Too much time sitting and staring at a box, or even walking around inside the house with a portable game player or some other portable electronic device, means the child is not in the real world mixing it up with warm-blooded humans. Spontaneous, face-to-face interactions with other children are not only necessary for the development of a solid self-concept, but there is even evidence that it improves thinking and intellectual abilities. Psychologist Oscar Ybarra found that more time talking with friends was correlated with higher scores on memory tests.[2]

According to some researchers, among them Chang-Hoan Cho and Hongsik John Cheon in the *Journal of Broadcasting & Electronic Media*, evidence suggests that children are more exposed to negative Internet content than their parents realize. This cuts across all levels of family income, parents' educational level, age of children, etc.[3] Leisure Satisfaction Scales (LSS) administered to 134 teenagers outside a cyber cafe showed that "*Web surfing frequency and life satisfaction are negatively correlated....* A deep absorption in online gaming may result in low school grades, deterioration in interpersonal relationships" and serves as a way of avoiding problems. These results show that the time spent on games and online activity increases anxiety.[4]

Because of peer pressure, Garrison and his buddies feel compelled to add "friends" to their Facebooks even though they've never even written to these people, let alone met them in person. They're encouraged by the peer group and social networks to share more information than they feel comfortable sharing. This could include membership in fan clubs, invitations to join causes, videos and games. The average Facebook user has 100 names. These naïve souls are sharing personal details with casual acquaintances or people they don't know that may have questionable or even criminal backgrounds.[5]

Privacy, "sexting" and cyberbullying.

Is Facebook growing up too fast? Facebook members are now challenging the company's right to license, copy and disseminate members' content worldwide. In a *New York Times* article, journalist Brad Stone believes, "People, of course, sometimes like to keep secrets and maintain separate social realms—or at least a modicum of their privacy. But Facebook at almost 200 million members is a force that reinvents and tears at such boundaries."[6]

Mizuko Ito's research raises the prospect that issues of privacy may have an even greater impact on teens. When kids try to impress others in their Facebook, or elsewhere, by sending in worldly, sexy images and comments, they don't always realize these profiles are public and permanent and travel beyond expected audiences. They may be difficult to eradicate after-the-fact. This presents challenges for both adults and children. Of course, this has been dramatically expanded via internet phones and "*sexting*." (Sexting is the act of sending explicit messages or photos electronically). Parents tend to be ignorant about what their kids are doing online.

Halvin, a Hunter, sometimes gets a little frisky. He took a cell phone picture of himself while taking a shower and sent it to a few of his "girlfriends." A parent discovered it and Halvin was kicked out of his highly regarded boutique, public magnet school. The police take this seriously. Students like Halvin can be prosecuted. In discussing this issue in a separate case, Andrea Davis, a police spokesman, said in a *Tampa Tribune* article, "It's like rape. If it's a consensual exchange, fine. But if it's obscene and you're offended by it, it's absolutely possible to press charges."[7]

How do we define "sexting?" Is it something like pornography, where folks think they know it if they see it? According to a *Wall Street Journal* article, a District Attorney in Pennsylvania believed he was taking an understanding approach when he gave students involved in "sexting" an opportunity to take a special class rather than face legal charges. But the legal charge would have been pornography and some teens had nothing to do with sending their photos to others. One parent called child pornography "the nuclear weapon of sex charges." Fourteen students enrolled in the special class but three of the parents and the ACLU are suing the district attorney.[8] Another IT wrinkle that needs some attention? You bet.

And it's easy to delete "friends." Most sites allow removal of these ghostly acquaintances with a click or two—and without letting them know they've been dumped. One day you're on your friend's MySpace top ten list and the next minute you're not. How would you have handled deletion as a teenager? According to a recent study reported by a senior editor of *The New Atlantis*, Christine Rosen, in *The Wall Street Journal*, "...Americans now have far fewer close confidants than they did 20 years ago."[9]

"You are so fat. Your mama's fat. You retarded B...., I'm going to f... you up." What about cyberbullying? Journalist Donna Winchester reported in the *St. Petersburg Times* that it's on the rise. According to the Florida Department of Education, 18 percent of students in grades six through eight said they'd been cyberbullied at least once in the preceding two months. Jan Urbanski, Pinellas County supervisor for Safety and Drug-Free schools said, "It runs rampant in middle school, and continues to haunt kids through high school." School Resource Officers like Christopher Burke don't need statistics to tell them it's a problem. He says, "The threats and language would curl your hair. Kids are doing all this in the safety of their own homes."[10]

Things can get nasty at the college level, as well. According to an article in the *St. Petersburg Times* by Alexandra Zayas, one site became notorious when it created forums for 500 schools, encouraging students to spill dirt on one another. "Common threads pondered such questions as, 'Who is gay? Who snorts coke? *Who has a sexually transmitted disease?*' Concerns about suicide, stalking and sexual predators created problems for this site's business, but apparently the economy brought the site down. One student government president said, "Good riddance."[11]

Journalist Jessica E. Vascellaro writes in a *Wall Street Journal* article that when teens want to "friend" someone new, they worry about whom to friend and whether their own friend requests will be accepted or ignored. Lingering in cyberspace is no fun, but face-to-face contact with old friends and possible new friends is not required. No need to explain the change in circumstances or negotiate or learn about whatever might have occurred to change someone's status. There are two sides to every story, but the machine makes it nice and easy to ignore that axiom. [12]

Friendships are often fleeting. "Click, click and you're outta here." At least Garrison doesn't experience separation anxiety, but is that good? In my opinion, *no* anxiety means *no* friendship. Just one machine logging on to another machine...then logging off.

Maybe for students, this is just another version of "hooking up," and that's also fraught with risk. Just hook and unhook, like plugging a machine into a wall socket.

Someone once observed that the original instant message was "A kiss on the lips." There's definitely some truth to that statement. One study of undergraduate students compares face-to-face relationships (not kissing) with Internet chat programs. The findings indicate that face-to-face groups feel more satisfied with their experience and enjoyed higher degrees of closeness and self-disclosure with their partners. This is hardly surprising to older generations, but might surprise the Digital Natives.

Recently, a college senior reflected on her life with electronics. In a *Notre Dame Magazine* article by senior Lourdes Long, she thought she might be addicted when she found herself checking her e-mail at stoplights and she realized that practicing the art of walking and e-mailing *at the same time* was harder than just walking and talking. She believes her quick messaging helps her as a student leader, but she does have concern that she'll "...miss a chance for Eye Contact, walking down the quad reading *The New York Times* on the tiny screen." She states, "In a culture structured around the use of these electronics, an inability to keep up with the rapid evolution of technology can feel isolating."[13]

As discussed earlier, the inflated ego resulting from IT often follows the young person into college and beyond. One college intern with whom I spoke worked at a bank, and when her supervising teacher asked her about the experience, the student said she didn't really like the company and would never think of working there. When the professor asked what was wrong, the student said the boss was not especially friendly and didn't thank her for her hard work. The professor had to let her know this was not an unusual experience and that the boss's job was to ensure prosperity for his company and maintain jobs for his workers; not to please or impress a college intern.

What this Digital Native doesn't realize is that she has to drop the customer mentality. Her bosses don't operate like the electronic games industry, where the customer is king. The inflated self-concept mantra has existed for some time now, but the IT experience is likely to increase the level of undeserved entitlement.

Do online relationships give the introverted (usually Gatherers) an opportunity to make friends online? In the article "The relationship between unwillingness-to-communicate and students' Facebook use" by Pavica Sheldon, a survey of 172 college students seemed to support the "rich-get-richer hypothesis, which states that the Internet primarily benefits extroverted individuals." They had significantly more Facebook friends than socially anxious individuals. Introverted teens did not develop more friendships, but access to the site helped them with feelings of loneliness, they said.[14]

In the article "Expression of Identity Online" by S.R. Stern, published in the *Journal of Broadcasting and Electronic Media*, self-disclosure is one of many markers of mental health. Real friends can disclose personal feelings and attitudes—*in person*. This is both a measure of a mature ego and a way to make the ego stronger and more flexible.[15] Sharing information about hobbies and rock groups is just too superficial to achieve true attachment—and, sorry, sharing through avatars and parallel worlds doesn't work either.

Garrison reveals a few things from his private self that he shouldn't, but mostly he's keeping his thoughts and emotions to himself. Research shows that young people are unlikely to disclose their innermost feelings, insecurities and anti-social tendencies online.

Writers who believe the machine culture helps thirteen-year-old Garrison build an identity are violating the *Developmental Principle*. This principle simply states that *children are not midget adults*. They're not little grown-ups, using the wondrous machine era to add new and interesting facets to their solid and intact identities.[16] The frontal lobes of the brain, for example, don't complete maturation until the age of eighteen or even later. These areas of the brain handle impulsivity and integration of thought. They pull together disparate thoughts and emotions to provide unified expression and behavior.[17]

Thousands of tiny bits of unstructured and unrelated information won't build Garrison an ego, but they can add to his confusion. Garrison might be able to reinvent his *public* self on MySpace or create an interesting avatar, but all this could lead to further splintering and confusion. Garrison isn't a kid in a thirty-year-old body whose main problems are from the *outside in*, in the form of Internet violence and Internet predators. Garrison is a thirteen-year-old whose identity is in jeopardy from the *inside out*. That's not to ignore outside threats, which are real, but first we need to help Garrison build a human identity and fend off those dangers.

Of course we wouldn't dare restrict the use of Garrison's many machines, because we don't believe in censorship. Do we? No, we shouldn't make him feel different from his friends who have every conceivable machine and electronic game, according to Garrison. And we hope that with the aid of technology, his mind will be opened and his unlimited potential reached. Okay, I'll admit that we might have to use a little censorship, but we'll call it something else, just to be politically correct—i.e. reason, judgment and the *Developmental Principle.*

Actually, we censor children from the time they're born—and even before they're born. We implore pregnant mothers not to smoke because of possible damage to their children and we restrict youngsters from tobacco and alcohol and make them wait until age sixteen to drive an automobile. Of course, they want to drive at earlier ages. But instead of letting them hit the road, we take them to such places as amusement parks, where they can navigate specially equipped racecars with low centers of gravity that are safe.

That safety comes in large part because they are *on a track* which guides them and gives them direction. Parents need to track their kids by providing structure at home. We're even careful about how much water infants get. Too much water can kill infants. And for sure, we don't let our children play in the street.

In 1960, the hippies knew that young people had splintered self-concepts. Comedian Bill Cosby states unequivocally that all young people are brain-damaged—and he's right, of course. It's difficult for kids to build a strong identity. Garrison needs structure, values and role models—not make-believe friends, mystical avatars and bleary-eyed parents slumped over their machines.

Precocious kids.

What about the adolescent culture itself? Today it mostly begins not at thirteen as it used to but at age eleven. Will early exposure to a wide array of enticements and manipulations by adult marketers push this age down even further? Now, starting at the age of four and five, some parents (especially those wanting to enter their daughters in beauty contests and pageants) parade them around in adult makeup and fashionable clothes.

Marketers do their best to entice children to nag their parents to buy their products. Sugar-filled cereal boxes in brightly-colored packages are placed at strategically low levels in grocery store aisles. This way the little ones can see them and scream for their favorite bright box and cartoon hero. The difference is that parents see what their young children see, whether that's cereal boxes or Saturday morning cartoons on television.

Too many kids have televisions in their bedrooms or play video games without parental supervision. Many parents, especially Hunter types, are less computer-savvy than their children and have no idea what their children are seeing *or feeling* while playing a game or visiting Facebook or MySpace. And unlike parental advisory ratings on content of movies and electronic games, it's process is that of concern. What are the machines doing to kids' brains, their time and their relationships?

Another example of parents not knowing where their kids are, even when they're right at home, is the case of Ginger, a sixteen-year-old Gatherer. She needs to do well on her final exam in History which is scheduled for the next day. Her parents insist that she go to bed early in order to be alert and fresh for the test. She's also eager to do well and retires at 10:30 P.M. Ginger is enjoying a restful sleep when, at 2:00 A.M., her cell phone vibrates. She spends the next hour-and-a-half counseling a friend who is having problems with her boyfriend. Her parents don't realize that *Ginger leaves her cell phone on all night, every night.* The next day, in a bleary-eyed state, she drops two letter grades on the big test.

Do you hear me?

Some of you may think you can communicate over e-mail as effectively as, or even more effectively than, in person-to-person situations. In the article "Egocentrism Over E-Mail: Can We Communicate as Well as We Think?" published in the *Journal of*

Personality and Social Psychology, five recent experiments showed that it is very hard to communicate without nonverbal "cues such as gesture, emphasis and intonation." People are overly confident and the researchers believe this may result from "egocentrism, the inherent difficulty of detaching oneself from one's own perspective."[18] You may think that others are receiving or understanding the statements you're writing in exactly the way you intended them—for example, as funny or sarcastic—but your electronic audience may not be interpreting them that way. Again, no huge surprise here.

This one's not just for the teen driver; it applies to adults as well. We've already talked about the dangers of using a cell phone while driving, but one study compared cell phone usage with passenger conversations, according to the article "Passenger and Cell Phone Conversations in Simulated Driving," published in the *Journal of Experimental Psychology.* The number of driving errors was highest in the cell phone condition. Passenger conversations differ from cell phone conversations because the surrounding traffic becomes a topic of conversation which actually helps the driver. Situational awareness increases, while the complexity of speech between the driver and passenger is reduced as traffic increases.[19]

Not only does the use of cell phones while driving lead to auto accidents, it also create problems *after* accidents. People used to just stand back and gawk when they saw car accidents. Now, with cell phone cameras and text messaging, the news spreads and hundreds of people may converge on the accident scene. This hampers investigations and rescue efforts. In one accident, emergency crews couldn't make needed calls. The massive amount of phone calls and texting had locked up the airways.[20]

First steps.

Now, let's take a look at the innocent ones whose brains are most vulnerable. One study found that 75 percent of kids in the United States, ages one to six, watch television and 32 percent watch videos for an average of an hour and twenty minutes a day. According to the Kaiser Family Foundation, more than one-third of three- to six-year-olds have television sets in their bedrooms.[21] Children spending time watching television without their parents is related to significant reductions in time spent interacting with parents[22], and background television interferes with toddlers' ability to focus on play.[23] There is a correlation between early television exposure and subsequent attentional problems in children[24], and television viewing stimulates an area of the brain that could put children at risk for future sleep disorders.[25]

In an article by Thomas N. Robinson, the effects of exposure to television and video games on aggressive behavior and attitudes were studied. Third and fourth grade students in one school were exposed to a six month curriculum designed to reduce television and video game use. Parents were encouraged to help their kids stay within a viewing budget. After eight months, these children were reported by peers to be less aggressive at home and in school when compared to the control group.[26]

One vivacious six-year-old girl with whom I talked was happy to share her IT experiences. She had definitely picked up the lingo. "You can play multi-person or against the computer." She told me about her favorite Web site where she plays a game "...and you knock out people and not get yourself knocked out. You use bombs and shells, and the red shell tracks you." Her eyes got wider. "You can't get it off." She also reported that her older brothers could win money on the Internet.

Money trap.

Some ten- to fourteen-year-olds also mentioned money while discussing their forays into computer land. They said they could enter a fantasy sports site and "even win money." It turns out they weren't sure how to do this and had never won any money, but it seemed to catch their attention. I went on a Web site offering free games. As soon as I got on, I was informed that I was the 999,999 visitor and had won a gift card worth $1,000.00. Meanwhile, a picture opened of a girl's face and I was told she had an I.Q. of 105. "Can you beat it?" the anonymous screen asked.

Then I was bombarded with pictures of three popular actors whose faces even *I* knew and asked to identify them, followed by an offer for a $1,000 scratch-off card. A psychedelic pink sign kept flashing, "Continue, continue" and I was told about a daily prize winner. While all this was going on, I learned that my lucky gift had a few minor conditions, like possibly purchasing products, applying for a loan, credit card opportunities and transferring my bank balance within 180 days. Would it ever end?

Dangers of video games.

Let's look at some of the more obvious dangers out there. In a two-year longitudinal study, according to the *Journal of Media Psychology: Theories, Methods and Applications*, Dr. Werner Hopf and his colleagues found that the more frequently children view horror and violence films during childhood and the more frequently they play violent electronic games at the beginning of adolescence, the greater their violence and delinquency will be at the age of fourteen.[27] Psychologists Craig A. Anderson, Ph.D., and Karen E. Dill, Ph.D., found in the *Journal of Personality and Social Psychology* that "even brief exposure to violent video games can temporarily increase aggressive behavior *in all types of participants*."[28] (my italics) Electronic games are more harmful than television and movies because they are interactive, very engrossing and require the player to identify with the aggressor.

Maybe we should restrict the sale of violent video games to minors. A U.S. Circuit Court of Appeals *struck down* a California law *banning* the sale of violent games to minors and cited the video game industry's voluntary rating system. Did the court take the *process* into consideration or just the *content*? I'd say that leaves it up to parents, but can parents get their heads out of their own machines long enough to see what their children are up to?

Fat machines?

New Democratic Party Member for Parliament Louise Hardy studied 2,750 eleven- to fifteen-year-olds and found that averaging more than two hours a day of television or computer time leads to poor performance on fitness tests. And those who played more entertainment games did poorer in school and had a greater risk for obesity.[29] In more than 173 studies conducted since 1980, three-quarters of them found that media viewing was associated with negative health outcomes.[30]

When children watch more than eight hours of television a day, it is a likely predictor of later obesity. No TV the first two years of life and less than an hour per day until the age of six is recommended. More than two hours per day exposure to television and/or electronic games significantly increases children's risk for asthma, according to an NBC News report.[31]

And when kids go outside to run off some of those calories, extreme Gatherer thinking, regulations and lawyers are sending them back inside. Syndicated columnist George Will discussed Philip K. Howard's book *Life Without Lawyers*, stating "Because of fears of such liabilities, all over America playgrounds have been stripped of the equipment that made them fun. So now in front of televisions and computer terminals sit millions of obese children, casualties of what...Howard calls 'a bubble wrap approach to child rearing' produced by the 'cult of safety.'"[32] So, we've got a machine inside and a bubble wrap outside.

Obesity can lead to heart disease. In the article "Studies find clogged arteries in obese kids," Dr. Geetha Raghuveer reported at the American Heart Association Meeting in New Orleans that "the arteries of many obese children and teenagers are as thick and stiff as those of forty-five-year-olds....'As the old saying goes, you're as old as your arteries are,' she said. 'This is a wakeup call.'" And research is showing that "fat kids become fat adults."[33]

Stopping Children's machine addiction.

Here are some of the things to look for if you're concerned about your child's electronic addiction: Ginger seems obsessed thinking about online activity and constantly anticipates the next online session. Heather's unable to cut back or stop online activities and longer time periods online are needed to give her a feeling of satisfaction. When her parents decreased her video game and Internet access, it led to restlessness and irritability. Harvey tries to conceal Internet activities. It's beginning to interfere with his school and social relationships to a significant degree and now provides an escape from his real problems.

Studies have shown that self-reactive incentives take over in game play. Relief of boredom, lessening loneliness, killing time and pure escape can lead to a loss of self-control over consumption and can trigger repeated patterns of game playing. [34]

Internet addicts, and even those who have become habituated or are just intense players, tend to feel the pleasure even before they boot up their computer. Even sitting in the same chair or being close to the computer may get the adrenalin started. Some of our students have told their teachers that the sounds made by the dial up connection affect them strongly. They're not sure how it affects them, but it is having an impact. This is similar to psychologist Ivan Pavlov's research on classical conditioning.

When Pavlov's dog smelled food and began to salivate, he rang a bell. Later, the ringing bell alone, without the presence of food, resulted in salivation. And after dogs—and people—are hooked, you don't need the food or the game. The bell and the dial-up noise will get things rolling.

If you're a smart parent and restrict game and television time, observe your child's behavior prior to her game time. Does he or she feel good just sitting in the chair near a video game system? Does your child get antsy just before it's time to start the game—kind of like a drug addict just before getting a fix? No, it doesn't mean he or she is addicted, but it's a testimony to the machine's power and the need for parental guidance.

Here's a scary question: If kids spend a lot of time gaming, at what age will they no longer choose arts, crafts or outdoor activities when offered to them? Age nine? Age twelve? Age four?

Ten Cyber Commandments for Teens

This list of ten positive teen values and beliefs comes from a brief survey of sixteen-year-old girls in April 2009.

1. ggr8—God is great.
2. ssmt—Science is smart.
3. fmfst—Family first.
4. ily—I love you.
5. Sxw8—Sex can wait.
6. khuf—Know how you feel.
7. drdd—Drugs are dumb and dangerous.
8. ryb—Read your Bible.
9. sc%l—School is cool.
10. mgb—May God bless.

Top Twenty-One Instant Message Codes

Can you decode your teenager's instant messages? Here's a list of twenty-one commonly used instant messages and seven smileys. This list comes from a survey of sixteen-year-old girls in Bradenton, Florida, April 2009.

1. wu—What's up?
2. g2g—Got to go.
3. w/e—Whatever.
4. b/c—Because.
5. rofl—Rolling on the floor laughing.
6. btw—By the way.
7. ttyl—Talk to you later.
8. ily—I love you.
9. bbl—Be back later.
10. lmao—Laughing my *#!* off.
11. brb—Be right back.
12. bff—Best friends forever.
13. omg—Oh my god.
14. asl—Age, sex, location?
15. l8r—Later.
16. rly—Really.
17. nvm—Never mind.
18. ty—Thank you.
19. kewl—Cool.
20. 2nite—Tonight.
21. 2moro—Tomorrow.

Smileys:
1. 8-) glasses
2. 0:) halo=angel
3. ;) wink
4. :P tongue out
5. >:) evil grin
6. :D laughter
7. :O yell

Ten Useful Tips for Parents

1. Build in time for family fun and personal communication first; then leave time for other activities on your weekly schedule.
2. Follow pediatric guidelines for electronic game and television usage. Limit five- through eight-year-olds to thirty minutes of electronic game time per day.
3. At family and other gatherings, place a cell phone box (Hot Box) at the front door for temporary deposit of all cell phones.
4. *Lecture* your kids about exposing themselves on Facebook and other sites. Other people now own and control their postings. Think privacy. Think bad reputation. Popularity is a myth. A few good friends will see them through.
5. Build in family game time—electronic and otherwise.
6. Build in fun reading time: even fifteen minutes per day is a start.
7. No instant messaging at the dinner table, when eating out or during study time. Keep cell phones in the Hot Box, *especially at bedtime.*
8. No cell phone use when driving. This means you, too, Mom.
9. No electronic game time immediately after discipline. Don't reward bad behavior.
10. Know what electronic games your child is playing. Think of *how long* they're on. Cut the violent ones.

Three Useful Tips for Kids

1. Tell your parents about cyberbullies.
2. If you're scared or confused talk to your parents, school counselor or other trusted friends and adults.
3. Relax. You're not an adult. It will all come together later—even if your friends think you're not cool.

Help for Parents

These companies assist responsible parents in providing safe environment for their children. The *Developmental Principle* comes into play here. Children are not tiny adults. They require varying degrees of structure depending on their ages and maturity.

These programs should be introduced by talking with your child. They offer a good opportunity for discussions about trust, obedience and respect for the family culture.

Yoursphere.com is a social networking site that limits membership to kids. Adults are denied entrance.

Websafety.com sells software for downloading to your child's cell phone and computer. It alerts you to the sending of inappropriate texts or photos.

Safe Eyes is a software product that lets you track your children's instant messaging, monitor social networking sites and impose limits on online minutes.

Illustration courtesy of Dave Coverly

"'2morro'...Could you use it in an incomplete sentence, please?"

7

Education, Relaxation and Games

Author Mark Twain wryly observed: "I never let school interfere with my education." Teachers, school administrators and the general public seem to agree that American schools are failing our kids. What can be done? IT? More money? Charters? Vouchers? Smaller classes? Stringent testing?

In 2005, the Federation of American Scientists, the Entertainment Software Association and the National Science Foundation convened a national summit on educational games. The consensus among participants was that many features of digital games could be applied to the increasing demand for high-quality education. They opined that "video game developers have *instinctively* implemented many of the common axioms of learning scientists." (my italics) Video games require players to master skills needed by employers, but unfortunately "testing programs fail to assess these types of skills," they concluded. They made the point that there are differences between entertainment games and educational games, and that "*educational games must be built on the science of learning.*" (my italics) They called for more research and lamented the fact that high design costs "make developing complex...learning games too risky for the video game and educational materials industries."[1]

Panel discussions produced some twenty-four attributes for the design of educational games. A number of the attributes were carried over from video games—no doubt about that. These included "user centricity", "novelty", "rewards", "enticing", "cool factor", "immediate feedback" and "user assistance, but not heavy-handed assistance".[2]

How long would these games retain their novelty and "cool" factor if introduced wholesale into classrooms? If we analyze more closely, it seems as though some of these individuals reluctantly agree that video games are not educational games and are not designed to teach a desired learning outcome. As we discussed previously, they're going to be enormously expensive to produce and much research is still needed.

I don't think information technology is the solution to our school problems at this time. Robotic, mechanical thinking is causing enough problems. If we keep our eyes on the developing child and not scintillating graphics, IT might eventually become another contributor to a well-rounded education.

What does the machine process teach us, anyway? Immediate rewards make electronic games interesting, but they reduce frustration tolerance and the ability to delay gratification. They may increase impulsivity and inattention, resulting in boredom in school. Machine thinking promotes the use of objective (true-false and multiple choice) tests at the expense of subjective (essays and writing skills) assessments.

Imagination and creativity are buried. The basic code level games *do* hammer away at a couple of useful accomplishments, however. Fine motor coordination and directionality skills might help our kids sort buttons on an assembly line some day. According to a *Wall Street Journal* article by Sue Shellenbarger, a better approach is that used in some Head Start classes today. "Tools of the Mind" is a curriculum that attempts to train "executive functions" through role-playing exercises where children help each other to learn about everyday experiences. It claims not only to improve creativity and mental ability but also to improve behavior and to reduce attention problems.[3]

So far, we've looked at quite a bit of psychological research, but what do Digital Natives look like at ground zero, where the action is? Direct observation doesn't give us the reliability and precision of elegant statistical applications with such "household" names as Latin Squares, multivariate, factor analysis and the ever popular Mann-Whitney U and Chi-square. Simple observation has yielded good results and even breakthroughs in science; although in recent years Hunter-style "eyeballing" has been replaced by the MRI and other machines. The insights and diagnoses of great men and women must give way to the incremental approaches of the Gatherer scientist, or so the Gatherers say.

For example, A.R. Luria, the Russian neuropsychologist, helped us understand frontal lobe function by studying the impact of gunshot wounds to the head at the Russian front during World War II in his book *The Man with a Shattered World*.[4] This research is still applicable today and gives us a better understanding of the executive function of the frontal lobes in children's learning.

Setbacks in motion.

What are highly trained educational clinicians and teachers seeing in classrooms today? A variety of problems:

- VISUALIZATION: It's helpful to look at middle school kids because they've been exposed to advanced electronic games over the past six years. What do we find? For one thing, they can't visualize. Wouldn't you think those hundreds of hours looking at gorgeous graphics would improve visualization? The research already told us that kids improve their peripheral vision and show faster response times to flashing lights or letters on a screen as well as their ability to assemble flat puzzle pieces to complete a design. Logic says they should be excellent visualizers.

  But research has also told us that these quick responses aren't activating the frontal lobes. What we're finding in our classrooms are kids who can call out words (decoding), but who can't visualize what they read. This important visualization process requires imagination, creativity and frontal lobe functioning. What to do? According to Patricia Lambert, Principal of the Center Academy School in Pinellas Park, Florida, instructors in the school are now teaching *middle school students* to visualize what they read! Among other techniques, teachers read paragraphs to the students while the students draw pictures about story content. We also give them pictures of blank faces and read descriptions of their appearance. Our students then draw expressions on the faces."[5]

And our middle-school kids are passive learners. Oh, so passive. They're waiting for something to happen. Like obedient servants, they wait for "the game" to start and to be told what to do. People who think electronic games are the passport to independent thinking and educational reform need to come to grips with what's really going on.

- DEPRESSED AND ANXIOUS STUDENTS. If kids play electronic games before school or at lunchtime, they're drained when the next class starts. Is it helping them reduce stress the way playing outside on green grass does? No. If they play touch football outside they register normal fatigue. They flop on the playground, laughing and talking. After the electronic games they remind one of deflated balloons...and they're edgy and irritable. They have many of the signs of *continuous partial attention,* which was discussed earlier.

- SHORTER ATTENTION SPANS. They can't read silently for three or four minutes, and these are not kids with attention deficit/hyperactivity disorder (ADHD). If the class shows an instructional video, they can't watch it. They take a bathroom break or find some other diversion rather than endure what, for most of us, is an interesting audio-visual story. Related to this is their lack of frustration tolerance. They will slam their mouses down in fury when they struggle on computers. Yet those same kids observed making bad plays during basketball games, do not react as quickly and explosively.

- VERBAL FLUENCY. Very few of these kids speak in complete sentences. They can parrot information, but many can't organize and initiate it. They lack fluency in reading, speaking and writing. They'd much rather *text* than talk. Teachers have noted a loss of listening skills as well. This problem is logical. When kids spend all their time mesmerized by the "machine" and its visual programming, they're not doing much listening, talking or socializing. Even our more verbal Gatherer types are impacted.

- MULTITASKING. Yes, just like their parents. If you watch them, from the outside, it's pretty impressive. Computers, MP3 players, texting and more—all at the same time! But look closer and you'll find that they're skimming the surface. Little is going into their brains and little will be remembered. And they always choose computer games first—*always*. Texting and the rest of it take a backseat—far, far in the back. It's becoming more difficult to motivate kids to use even the most interesting educational programs.

- COMPUTER DEFICIENCY. Okay, so there may be real drawbacks to electronic games, but at least they sharpen kids' skills on computers and prepare them for the modern workplace. Right? Not so fast. Our teachers have seen a significant drop in computer mastery in the past five years. Five years ago, middle school kids had mastered Microsoft Word, PowerPoint and spreadsheets. Now, about 50 percent have mastered PowerPoint and 30 percent can do spreadsheets. They know game strategies backward and forward, but can't use such programs as Excel. Is this related to video games? When you see kids use two fingers on the keyboard, just as with "the game," it makes you wonder.

- CAN'T GENERALIZE. Compared to middle school students a few years ago, today's students seem narrowly focused and have difficulty generalizing. Inability to generalize is an extreme Gatherer trait and it means fewer opportunities to "think outside the box." A very bright student who found a reference to Roman political and military leader Julius Caesar, asked his teacher: "Is this Julius Caesar the same one we had in world history?"

These are teacher observations. With almost all the research results cited in this book, we need more and better research. As indicated earlier, we need replications of these studies by social scientists in separate institutions and, of course, without *any* connection to the "Gaming Industry," cell phone manufacturers, computer companies, etc. And more *longitudinal* studies that follow very young children to maturity are necessary.

Filling in the dots.

Many educational professionals believe that objective testing and machine-scoring permit more frequent assessments of progress and serve as a guide to overall school improvement. Some teachers are not so sure. They fear we will teach to the test rather than to the child. Here's an example of the tail (the machine) wagging the dog (your child and his or her school). Let's say you're a school principal. The machine determines that you need to have high test scores for your school to be recognized and receive federal monies (and for you to keep your job).

Along comes company X and it says: "We can raise your school test scores and keep you squarely in the government trough."

"How can you do that?" the naïve principal asks.

"We not only know the questions on this 'secured' test of yours; we also know the importance and statistical weighting of each item on that test. Young Garrison, over there, needs to learn sections F and G of the test before March, but our analysis shows that he probably can't do it. Harvey only needs to improve on sections L and M and our statistical analysis shows that he might make it.

"So, here's the deal, Mr. Principal: Use the monies allocated for young Garrison and put them at the service of Harvey. Give Harvey the extra tutors and curriculum materials and *let Garrison find his own way* and when his parents ask why he's not progressing—well, that's your problem. Too bad, nice kid, but raising the school's score is paramount and worth the sacrifice. Maybe we can offer Garrison some remedial work next summer to help him catch up. It's the only thing to do." Does this make sense?

There has always been a tug-of-war between Gatherers and Hunters in schools. Peter Walker, principal of the Evelyn Grace Academy in Brixton, in the RSA journal (Royal Society for the Encouragement of Art, Manufacturers and Commerce) states, "My starting point is similar. A point of reference for me is Neil Postman's book *The End of Education*, where he says that there are two questions to be answered in education: the metaphysical question and the engineering question. As a nation we tend to spend most of our time addressing the engineering one, the means by which we might change things, rather than saying what, fundamentally, school is for."[6]

School daze.

Why is it the child who is punished when adults cannot work out their problems? We build huge facilities (most people mistakenly think school buildings are schools) that require a big number of students in order to fill them. We wanted to end discrimination against minorities and thought integration of schools would help. It seemed like a good idea at the time and the goals were honorable, but the unintended consequence?

Our precious children, who count on us for health, safety and leadership, are stuffed into metal machines called school buses and driven great distances to attend school far from the nurturance of home and neighborhood. Sometimes they are yanked out of bed while it is still dark and deprived of the sleep necessary for health and learning. This is especially true if they are allowed to stay up the night before to watch television or play video games.

The same might be said for the regimented, mechanical achievement tests. What parents and taxpayers need to know is whether the child's teachers are good ones. I believe we do not know how to ensure that a quality teacher is working with each child, so we make the child take a test and then make inferences about the school, not the teacher. It doesn't make any sense.

Ironically, the test may not tell us that much about the school or the teacher, because so many other factors influence the child's test scores. Even if those variables were controlled, the test itself is biased more toward the Gatherer than the Hunter. From the Hunter's perspective, the test page, filled with boxes, letters and words, looks and smells like a crossword puzzle or some other devilish thing concocted by Gatherers.

Maintaining a balance.

Some teachers, ever mindful of these differences, have introduced instruction for the kinesthetic learner. These Hunter children delight in the joy of movement and learn better through large-motor and fine-motor activities, along with touch. Hunter children may learn spelling by lying on their backs and forming the letters with their arms, for example. Two students might stand and extend their arms to create an imaginary bridge for other children to crawl through.

Does that mean schools should encourage Hunter children to learn largely through movement and kinesthetic feedback? Should they tolerate a chaotic classroom environment that distracts not only the Gatherers but also the other Hunters? I don't believe so. The question is this: How do we maintain the balance? We don't want to lose the Hunters' imagination, creativity and emotional aptitude, but in order to succeed, Hunters must adapt to the sequential learning environment. Because Hunters are visually inclined, computer learning—as opposed to electronic entertainment—can help them learn without corrupting their minds.

Because Hunter children learn better through visual rather than auditory modalities, perhaps we should teach them entirely through visual avenues. Already, computer companies are touting preschool instruction in reading using visual feedback. As a psychologist who operates schools for children with learning difficulties, I will happily admit that computers can assist in the learning process if the program is designed to fit specific needs.

The school programs at Center Academy use visual approaches to audio-phonic dyslexia. These programs use visual avenues to teach children who are weak in auditory processing and who have reading and spelling problems. One technique uses raised letters. The dyslexic student touches the letters with two fingers while reading the words along with the teacher. Theoretically, this permits the information to bypass the Gatherer side (left brain) and come through the Hunter side (right brain).

But the fact that remedial approaches help children with learning disabilities doesn't necessarily mean these approaches should be used uniformly with all children. Most experts agree that for reading and spelling to take root, phonics approaches involving sound-blending are critical. As indicated in chapter 4, this situation is analogous to some experiments in the medical field. Massive doses of a certain vitamin assisted patients who had experienced open heart surgery, but when vitamins were tested on the general population there was very little evidence that they were effective. The vast majority of people get proper nutritional levels through diet, exposure to the sun, etc.

There are plenty of Hunters out there who need visual approaches to learning. Some of them may even need to skip phonics instruction! They'll be lousy spellers for life…but that's a lot better than "falling through the cracks" and dropping out.

Who's on first?

Mizuko Ito and her fact-finding group took the point of view of young people on their own terms without looking at adult-driven goals. She and her group wanted to be "agnostic" about learning and just see what the kids were doing. They also conceived of public education as a broader agenda and not just something that happens in the classroom. They felt that youth learn better when given authority over their own learning and literacy.[7]

All of us want to improve the learning process and this is especially true for the Hunter, who tends to be highly creative. No doubt our Hunter students would like to turn the school system on its head and let the students teach the teachers. Obviously, the learning process must make sense to the young person. The much overused term "stakeholder" is now in vogue and reinforces the idea that we all have a stake in whatever happens in our homes, our schools and our businesses. That's true, but some individuals have a bigger stake than others, one that is based on education, hard work and contributions to their communities.

Center Academy schools have been teaching through an individualized method for forty years. Students are given the option to prioritize their individual programs and—lo and behold—most of them follow "Grandma's rule," which is *work before play.* They choose more difficult tasks before going on to the easier ones. But this is always with the help of teachers and teacher associates. The students' individualized programs are planned by professionals who have spent many years in the classroom, working in the trenches. (And the play the students get as a reward is real play, not electronic play.)

In some school districts, girls and boys are encouraged to think about vocational careers as early as middle school. I hope the professional soccer and football leagues of the future have room for many millions of athletes!

As with most other areas of social and psychological development, the child's authority over his/her own education increases gradually with age. Many students, even in college, are not sure of their area of specialization or interest. In fact, it's not uncommon for individuals to change careers in midlife, when they finally discover their niches. In my opinion, this authority must be earned. And it is earned based on sound instruction, self-discipline and brain maturation.

Ginger Graham, a thirteen-year-old Gatherer, loves school. Everything about school is consistent with her personality. She likes to take multiple-choice tests, memorize facts and use her laptop to take notes. Her social studies teacher, like most teachers, is also a Gatherer, and she constantly rewards Ginger for her excellent computer work and test taking skills. Ginger's behavior is usually consistent and predictable. This reduces stress for her teacher, making her job easier. After all, this Gatherer teacher has to spend much of her time trying to organize Hunter students who want to do things their own ways, outside of the established protocol of the classroom.

Lately, the teacher has noticed that Ginger is distracted at certain times of the day. What she doesn't know is that Ginger is texting her friends both in her classroom and elsewhere. She can do this without looking at the keyboard and keeps her phone in the pocket of her jeans. At other times during the day, she looks at pictures of boys. Some of the boys have supplied snapshots of themselves, flexing their muscles. Maybe this is one time when the IT culture helps the Gatherer to develop a few Hunter instincts!

Most of us think texting is used a few times a day—okay, maybe several times a day. When one public school resource officer was accused of misconduct with an eighteen-year-old student, records showed that the two had texted each other more than 5,000 times in five months. This included 176 exchanges on one day and 592 exchanges on another day, according to an article by Rita Farlow in the *St. Petersburg Times.*[8]

Ginger's brother, eight-year-old Harvey, is a Hunter. Harvey's not with the program. School reminds Harvey of an underground nuclear fallout shelter in one of his shooting games. School's so confining that sometimes he can barely breathe, not to mention that he's always in trouble for one thing or another. When he does math computation, he usually gets the correct answer, but is graded down after failing to show the steps to his solution. His teacher thinks he copied from Garth's paper; otherwise, how could he get the right answer without doing the steps and anyhow, he has to learn sequential thinking. "That's what school's all about," she says.

Harvey hates sequential steps. They slow him down and confuse him. *There are no "partitions" in little Harvey's mind.* He doesn't like phonics either, because he can't remember the sequence of all those sounds. When he writes a story, he gets into trouble for not using note cards.

He wonders whatever happened to art, music and physical education. When he finally gets a fifteen-minute recess, he runs wild and attacks the swings and monkey bars with all his might, but this often lands him in the principal's office. His teacher is afraid he'll fall on the hard asphalt surface and skin his knees. She talks to his parents about medication to make him less active. *What's so bad,* he wonders, *about skinning your knees? And anyway, why can't they have some grass to play on? It really seems to settle me down.*

Sometimes he frustrates his teacher because he blurts out the answer *before* she even asks the question. It's a mystery, but somehow, perhaps based on the teacher's personality, facial features, gestures and the context of the lecture, Harvey knows what the teacher's question will be and responds immediately. His Gatherer teacher doesn't like that one bit. Harvey and his classmates are asked to take notes on their computers over the holidays and share some adventures with their classmates. Harvey doesn't take any notes because he is too busy playing outside, but his classmates love his talk because he jokes around and has a great personality. His teacher just smiles and shakes her head.

All students in the school are required to sell raffle tickets for the purchase of school computers. Most of the kids ask their parents and neighbors to buy their raffle tickets, but Harvey makes a booth out of cardboard boxes and sets it up at a neighborhood shopping center. Harvey has three times more sales than his classmates, but this causes him even more trouble. He needs a license to sell at the shopping center, and besides, his principal says he might have been kidnapped and murdered (or worse).

Harvey starts to have doubts about the value of school. He figures that anyone who would make a kid get up at 5:30 A.M. to ride forever in a metal machine called a school bus couldn't be too smart. Unlike his parents and teachers, Harvey knows he's just a kid, not a grown-up, but he wants to please his parents, so he'll try to be a good student like his sister and stop coming up with crazy ideas that don't fit in. Creative, non-sequential thinking is what gets him in trouble. Maybe he'll just stop thinking.

Many boys rely on Hunter strategies from an early age. At age five, boys are significantly superior in spatial skills but lag behind girls in vocabulary and reading. Some people have concerns that girls aren't learning as much as boys because they have less interest in higher mathematics and social interests interfere with their studies.

But recent research shows it's not the girls who are left behind the academic eight ball—it's the boys. Guys suffer from more brain-related impediments to learning. They have significantly more learning disabilities and attentional problems, not to mention a whole array of nasty developmental disabilities.

Are boys really shortchanged in our schools? Hunter boys (and some Hunter girls) are sailors who've been set adrift in a linear world. The vast majority of special education students are boys and some observers think special education classes could be a form of discriminatory single-gender education.[9]

As discussed earlier, left-brain/right-brain research is the neural basis for the Gatherer and Hunter personalities. When neuropsychologists illuminated left-brain/right-brain differences, the ink wasn't even dry on the research reports before a bunch of public school individuals quietly murmured, "Where have we been? How can we use this stuff in the classroom?" After reviewing the initial research, some educators condemned the lack of right-brain stimulation in the classroom. If right-brained students showed underachievement, some educators believed this resulted from overdevelopment of the linear left brain (Gatherer) and under development of the right brain (Hunter).

Conventional education was blamed for teaching Gatherer analytic brain skills such as reading, writing and spelling at the expense of Hunter skills. Even the Gatherer teachers were suspect. Time to get them out of the classroom and bring in some Hunter types, the educators proclaimed. Fortunately, the teacher's union said: "I don't think so; not on our watch" and that ended the grand inquisition.

How do we know if a student—or a teacher, for that matter—is a Hunter or a Gatherer?

The solution should be fairly simple. Right? Let's put our heads together and come up with a test that taps into Hunter capabilities and strategies. How about requiring the assembly of puzzle pieces to make a picture? This is the skill that showed improvement when kids played an electronic puzzle game for eleven days. We'll place the pieces on the table in front of the child but in varying positions, right side up, upside down and at varying angles. We won't show the student a picture of the completed puzzle and we won't say anything except "Put these together." If the child does well, we'll know he's a Harvey or she's a Heather, right? They'll use Hunter strategies to look at the big picture and rely on abstract thinking and visual memory to *suddenly* apprehend the configurational whole; then quickly assemble the pieces. If not, we'll know he's a Garth and she's a Ginger.

If it were only that simple. These blessed children do have a mind (or two) of their own. What we forgot is this: It's the *process* or the *way* the material is approached and not just the content (visual or verbal) of the material. So what does Ginger do to destroy our neat little test? She moves the edges of the discrete pieces together in a sequential, linear fashion, with little awareness of the big picture, to discover where they best fit, and she does it quickly.

But, since we were alert, we noticed that Heather apprehended the big picture and put the puzzle together all at one time, while Ginger did it piecemeal. So our test still worked, but it wouldn't have worked if we'd just compared the time it took them to complete the puzzle or looked at a score on a computerized test sheet.

Now let's make up a test for Gatherers. We'll simply ask a series of questions to find out what they know. We'll start with easy questions like, "How are a dog and a lion alike?" and then go on to tougher ones, like, "How are a church and prayer alike?" Now, consider what we're dealing with this time. This test is all verbal and there's nothing to visualize or touch. No way Ginger or Garth can mess us up this time. Right?

But there's one little flaw here. As the questions get tougher, they become more abstract and more Hunter-like, and Captain Ginger may shout over to Captain Heather: "Hey, I need a little help over here. I got the ship into this snug little harbor, all right, but this universe stuff is a bit nebulous." Now, consider the meaning of this. The test questions changed from Gatherer to Hunter in midstream, and without a fair warning.

This change from concrete to abstract functioning is also seen in math and other subjects. Harvey solves abstract math problems without using sequential, step-by-step rules. In fact, most Harvey types are relieved when abstract math, such as algebra and trigonometry, are introduced into the curriculum.

Don't feel too badly about your test construction capabilities. Both of the above tests are similar to older editions of the Wechsler Intelligence Scale, the world's most widely used intelligence test. Surprised? In defense of this test, which is sometimes used by neuroscientists for research, it was not developed to tease out hemispheric function or personality type and new revisions remain the world standard for validity and reliability.

There are many other examples of individuals not doing what they think and say they do. I observed teachers at one school who claimed to use Montessori methods and I also observed teachers at a school that was billed as a center of creativity where students played musical instruments in a tree house to learn math. Both schools believed they were enhancing creativity through Hunter-type, experiential learning, but the process of rigid, sequential programming was far more linear than creative.

Many teachers select the teaching profession because it's consistent with their organized, sequential approach to life. In the '30s and 40s, whole-word methods were used for reading. Children looked at pictures and memorized the words based on visual configurations and context. This should have saved the Hunter kids. It helped a few, but even then, the stories were written to "promote *sequential* habits of learning," as observed by *Cathedral Basic Readers.*[10] The Hunter child was never turned loose to enjoy the pictures and create his or her own stories. If entertainment games can ever grow into learning games, perhaps the Hunter child could enjoy creativity *and* "Best Practices."

Some have opted for separate classes for boys and girls. Okay, this might help with differences in activity level and sociability, but it doesn't address the fact that Hunters and Gatherers are represented in *both* genders. And furthermore, female teachers often end up teaching the all-male classes. Let's get one thing clear from the get-go: Boys and girls alike have different personalities and hemispheric tendencies. What's the solution? Awareness of personality and brain differences is a critical point from which to start.

In an ideal world, boys *and girls* who are weak in left-brain capabilities would not be taught to read and write so early. Of course, this flies in the face of the group that insists on showing off "prodigies" at the ages of three, four and five. This early blitzkrieg may be more of a babysitting service than an educational endeavor, but—unfortunately for the children—this can also result in frustration, failure and burnout. For almost thirty years I have warned about the dangers of children growing up too fast.[11] And thanks to our new scanning machines, in-the-womb education will undoubtedly become the next *sine qua non* (essential action, condition or ingredient) for modern parenting!

Online studies.

Our present college applicants have not been exposed to the full flora of IT culture and gadgets, but they may suffer from occasional bouts of pandemic pestilence. Applicants are relying on computer mixing and matching to build impressive applications to elite colleges. This sometimes results in blatant errors, however. For example, some applicants have plugged in the name of the wrong college when answering the question: "Why are you applying here?" And because it's so easy to send applications through e-mails, some applicants bombard administrative offices with thank-you notes and other evidence of narrow, machine-like thinking. Many of the applications are coming across as stilted, engineered and robotic, according to the article "How Not to Get Into College: Submit a Robotic Application" by Sue Shellenbarger.[12]

Online study sites are now offering college students test answers and professors' old tests. One professor said his students' use of these sites forced him to lessen the weight of graded homework to 10 percent of the final grade from 30 percent in the past. Students even log on before choosing a class to research average grades by different professors teaching the same course.

A professor at a small liberal arts college found that students were texting one another so much prior to class and after class that they had little to talk about during class. She designed what she called a breakout class where students could wrench themselves from their virtual world for forty-five minutes to interact with one another. She told the students that this would be a fun experience and the students really seemed to enjoy the face-to-face interactions.

An architect was asked to teach an advanced class at a well-respected liberal arts college. When a student told him that he couldn't find biographical information on a certain architect, the teacher asked the student how he went about his research. Naturally, the student said he'd used the Internet. The instructor took him to a classroom window, pointed to a large, gray stone building at the end of the quad and suggested the student start there.

"What is it?" the student asked.

"It's the library," the architect replied.

"Oh, I've never been there," the student said.

A different college professor acquaintance of mine, Cecil Cheek of St. Petersburg College believes the latest crop of students want to have it *their way.* He described it as a "customer" mentality. The customer is always right and little effort is put forth to achieve goals. One student reported to the Dean that the professor had "scowled" at her. Cell phone use is also rampant. This professor has a rule that if a cell phone is left on and a call comes in, the offending student has a choice between losing grade points or standing and singing the first verse of his high school or college alma mater. More and more of his students are taking online courses, which he believes is a gravy train for colleges.[13]

So the landscape is changing. Do we adjust to it or head it off?

More Innoculation: Twelve Tips for Parents

1. Support art and music at your school and at your school board meetings.
2. Don't be satisfied with teaching by computers.
3. Prohibit use of cell phones in class.
4. Don't let objective tests and "teaching-to-the-test" immobilize your child's teacher—and infect your child's mind.
5. Support recesses twice a day—on real grass.
6. Support structure and discipline in your child's classroom. Chaos and creativity don't make good bedfellows.
7. Support small schools, small school districts, vouchers and charter schools. If your principal doesn't know your child's name, your child's school is too big.
8. Reward teachers who plan carefully and who provide structure and nurturance. Get bad teachers out of the classroom.
9. No electronic games at school. That includes lunchtime and especially recess.
10. Read your child a short story and have your child draw pictures about the story. This teaches *creative* visualization.
11. Tell or read your children bedtime stories and ask your children to create changes and add their own plot lines.
12. Help your children, especially your Hunter kids, to improve their computer skills.

How to Help Gatherers Learn

- Gatherers are auditory learners and when they learn to read they should use phonetic approaches (analyzing the sounds of letters and groups of letters to figure out or "decode" words).
- Even when writing, it helps Gatherers to talk to themselves and verbally plan the sentence they want to write.

How to Help Hunters Learn

- Hunters learn to read by looking, not listening. The "look-say" approach is most effective for this group.
- When learning writing, Hunters may need to go over what they write more than once. They may leave out a word or two or come up with unusual spellings. Most people create what they want to write by hearing it in their minds first, but learning by listening just isn't the Hunters' approach.
- Even in spelling, the Hunter must see the words and study them with his or her eyes. Flashcards were practically invented for teaching math to the Hunter.
- Scribbling, writing or underlining really helps Hunters learn. Even in non-academic settings, Hunters do much better if permitted to use a pen or pencil to take notes, draw diagrams and pictures or just plain scribble. It helps them get the information into their brains.

8

Robotic or Romantic?

Mechanization and Marriage

Dr. Gary looks over his attentive audience, mostly women, and offers his practiced, professional smile. They are usually the same, these audiences. Toting swollen shopping bags and wanting to be part of the "Big Apple," they bask in the glow of the famous television psychologist. Maybe, someday, they would make their way to the head of the class to sit with Dr. Gary.

Dr. Gary still feels the butterflies churning, even after a few years of daily shows, because, unlike his audience, he can't predict the behavior of his "matinee clients," who will soon appear from behind the backstage curtain. Sure, he's already met with Hal and Gail Gather-Hunter to review their story and knows they were carefully screened from a pool of hundreds.

But there is always that slim chance they'll overplay their parts and disrupt the rhythm of the show; or, worst of all, make Dr. Gary look bad. If they follow the director's loose script of suggestions too closely, they could come across as stilted and unnatural, but if they get too comfortable, anything could happen. And that is not good. He walks straight to his audience, careful to stop at the pink chalk mark scratched across the floor.

"It's time to change your lives," he exclaims, unveiling his trusted, "best friend" smile. "Charge those batteries and get ready for takeoff!" (Audience applause.)

Finally, it is showtime. Gail (a Gatherer) and her husband, Hal (a Hunter), are drifting apart. A stranger, the mechanization process, has invaded their family. Gail is relying on her machines and beginning to act like them. This turns Hal off. Hal refers to her computer as "the other man in the house." He suppresses his anger and reactive depression by watching sports on three other machines.

Gail sits on the edge of her chair and stares intently into Dr. Gary's face. Tears wet her eyelashes. She is pretty, but overweight. Her bright, silk dress is tasteful and conservative. Her nails are bitten down to the nubs. Hal is rounded, but muscular and has an engaging smile. He settles comfortably into his chair. After a nod in Dr. Gary's direction, he leans back, looks the audience over and peers at the television cameras, balcony and staging.

It seems that Gail has engaged in two extramarital affairs with fellow employees. She expresses dismay that she would do these things, because she loves her husband and her behavior violates her deepest core values. After some probing from Dr. Gary, she says she must have craved attention. Her intimate co-workers bought her lunch and one of them brought her flowers. She poured out her heart to these sympathetic men, she says, telling them about her inattentive husband and her stressful, unpleasant job.

According to Gail, Hal frequently forgets to do the chores she assigns him. He seems so absorbed in watching television sports and playing electronic games with their only son that he doesn't even hear her. He had less and less time for her and when he isn't working, sneaks off for golf games with his friends. Gail feels he must not love her.

Gail withdraws to her computer or sends instant messages to her friends. Her job provides little satisfaction. She began working as a secretary in a department store but was soon promoted to the purchasing department, where she selects and sorts merchandise for discount sales.

Hal gets to *his* bottom line in a flash. Gail's co-workers are buttering her up for sexual favors, pure and simple. Meanwhile, he is working extra hours to provide her with nice things. Hal admits he plays too many video games with their son, but feels tired and stressed after long hours at work. He works in a depressed area of town in an auto parts store and dislikes searching through the warehouse for minuscule parts for old cars that are often unavailable. His customers repair their cars in the store parking lot. Hal finds his job both boring and disgusting.

Dr. Gary gently probes Gail's childhood and discovers that her father, a bookkeeper, watched sports on television with her older brother while devoting little time to Gail. Gail wasn't that close to her mother, either. She had few real friends and retreated to her room to read romantic novels and complete crossword puzzles. Hal's parents were divorced when Hal was three years old, and he can't remember his father, who was a traveling salesman. His mother never remarried. She smothered, controlled and indulged her overweight son.

The show's director signals Dr. Gary. He has only three minutes for his summation. Do they still love each other? Yes. Do they want to work this through? Yes. Well then, Gail needs to stop her sexual acting out now or risk destroying her marriage. She needs to realize that Hal is not her father and that Hal fears she is trying to control him.

Hal needs to spend more time with Gail and listen to her feelings. He is not to sort out her problems and give advice, but rather show that he's always available and loves her more than anything or anyone. He needs to do household chores whenever asked. They need to date each other at least twice a month. Gail and Hal need counseling. Dr. Gary offers a referral. Will they go? Can they afford the cost? Do they have insurance that will cover it? Or will Dr. Gary's television show pay the bill? Yes. Dr. Gary sends them on their way to live happily ever after and the audience gives its blessing with sustained and energetic applause.

Personality styles.

Will Gail and Hal live happily ever after? Hal will try to do the chores whenever asked and spend more time with Gail. She will come out of her room and join Hal in adult conversation, but will they keep their promises to each other and to Dr. Gary? If they don't fully understand the causes of their behavior, will the changes last? Their behaviors may be symptoms of underlying childhood experiences, but they may also reflect their personality styles and the cold winds of technology that have swept into their home.

They received some reasonable advice, especially considering the artificiality and time constraints of Dr. Gary's entertainment format. But knowledge of personality and brain architecture could potentially add much to Dr. Gary's recommendations and the eventual outcome. *What Dr. Gary doesn't know is that he is trying to communicate with two uniquely different personalities that are masked from him and from each other.* In addition, Dr. Gary's brain personality represents a third variable in this tricky, three-way communication process.

It's as if Dr. Gary is speaking to strangers who do not share his culture and needs a guidebook to help him explore two foreign societies. How does a counselor from a rural state empathize with these two people and their disparate cultures? Dr. Gary needs a foreign language dictionary that covers brain personalities. Without it, his advice may fall on puzzled ears.

So what are Hal and Gail's personalities? Hal is an extreme Hunter, much like his salesman father, while Gail has many more features of a Gatherer, similar to her father and quite unlike her stylish mother. Hal loves watching sports on television and playing golf, because he is drawn to spatial movement and hands-on activities. He has always disliked reading and detests computers, although he loves violent video games. Ironically, it is unlikely that these attitudes have much to do with his parents' divorce or Hal's lack of a father-figure as a role model.

Machines and marriage.

After a few visits with the psychologist to whom Dr. Gary referred them, the underlying story began to unfold: After eleven happy years of marriage, Hal began to notice that Gail was spending more and more time on her computer and this was long before her sexual acting out. At first she used it for an hour or two a day, but more recently she had expanded her time to five and even seven hours, sometimes at one sitting. Hal wondered if this was similar to his old drinking problem. He needed to keep increasing his intake just to reach the same levels of unhappy intoxication.

Whenever Gail finally staggered into the living room after a long bout on the computer, she was irritable and seemed dazed. Hal had never heard of ***continuous partial attention***, but she was so busy keeping tabs on everyone and everything that she couldn't prioritize or make thoughtful decisions.

In Gary's opinion, she had always been somewhat of a control freak, and at first her computer had helped her organize her life. But now the computer was controlling her and she was showing a lot of stress. Even her posture seemed rigid, and she complained of backaches and stiff fingers. Hal asked the therapist: "Doesn't she know her body is trying to tell her something?"

Hal reported that Gail was becoming more demanding and aggressive. She expected immediate answers to her questions and had little time for discussion. When Hal was speaking with Gail in front of Gail's friends, Gail tended to interrupt him and complete his thought or idea. She might hear one word that interested her and jump in on that word rather than appreciating the rhythm and flow of the conversation. She had always talked a lot and used speech for control purposes, but the therapist explained that under stress we regress to old patterns of behavior, so Gail increased her verbal output even as she increased her e-mail messages.

Gail was proud of her ability to e-mail quickly and she thought she was a good communicator. But, as discussed earlier, without the benefits of nonverbal cues such as gesture, emphasis and intonation, studies have shown that e-mailers have an overconfidence that is born of egocentrism.[1] Even though Gail intended to add sarcasm in a humorous manner, her acquaintances didn't perceive it that way. Hal noticed that Gail was becoming more frustrated as she tried to unscramble her poorly conceived messages and apologize for mix-ups.

In the happy days of their marriage, Gail came home from work or a social event eager to bounce ideas off Hal. He had learned that she was not looking for specific answers to her questions and thoughts, but wanted to share her life by disclosing her feelings. Hal was proud of himself for learning not to ask specific, bottom-line questions, and he thought his understanding responses were cementing their relationship. Now Gail seemed to have little time for communicating in this general and relaxed way. She seemed to talk only to control and had little time because of her computer. Hal felt left out. Was he jealous of a machine?

Hal commented on speaking to a friend at a party. The friend was also a Hunter and there were sometimes pauses in their conversation of up to ten seconds while they reconsidered an idea or just organized their thoughts. When Gail observed this, she felt embarrassed and uncomfortable. Just a few seconds of silence caused her stress. How could they just stand there, with friends all around, not saying a single thing? Hal wondered if her need for control has increased or if it is just her increased pressure to speak and act quickly.

Another time they were in a store and Hal found a box of tea he especially liked, but the price label was missing. He continued looking at the box and checking for the price when Gail interrupted in an urgent and irritated manner. She pointed to a store clerk and said, "Go over there and ask the man what the price is." Hal felt like a six-year-old getting a note from Mommy. She was also critical when Hal used such clichés as "better late than never," "what goes around comes around" and "back to the drawing board." Hal liked clichés and saw them as a way to quickly summarize a subject and keep details from getting in the way of his story.

Gail seemed to be doing more and more of what she called multitasking. When she was in a New York City taxi getting ready to exit the cab, she was talking to the driver, speaking on her cell phone and retrieving money from her purse to pay her fare. She lost her wallet and credit cards when she inadvertently dropped them on the floor of the taxi and left the cab without them. When Hal brought up multitasking as a problem, she was furious.

She'd had some close calls driving, as well. Gail insisted on using her cell phone from the time she got behind the wheel until she arrived at her destination. Hal had read an article in *Parade Magazine* by Lyric Wallwork Winik indicating that any type of cell phone use, whether handheld or not, makes drivers four times as likely to suffer a serious automobile crash.[2] Gail texted Hal to mind his own business.

Planning is another problem. Gail relied on her cell phone for last minute information. As a result, she didn't feel the need to plan ahead. Over time she lost the ability to plan effectively. She is always calling others for advice on simple matters and couldn't make decisions on her own.

Gail saw an advertisement for a riding stable on the Internet and thought it might do her some good. Unfortunately, she caught the wranglers smirking and shaking their heads when she tried to mount her horse. It wasn't because she was overweight and out of shape, it was simply that the cowboys had never seen a machine mount a horse before. It just didn't seem natural. Hal thinks a vacation might help. Gail agrees and they visited a historic city at the foot of a mountain. Even on vacation, their differences did not disappear. Gail wanted to attend lectures and visit every landmark possible, chalking them up as if she was doing an inventory. Hal wanted to go mountain climbing and get close to the earth. He was looking for a little adventure and wanted to enjoy the nightlife.

Gail continued to reply to Internet messages on the spot, rather than thinking through her responses. Her friends do the same thing. One time she didn't check her mailbox for a couple of hours while shopping and was berated by a friend for not getting answered immediately. Gail often signed off on requests with an ASAP, even when the item wasn't that important or timely. Hal thought she might cut back on television and newspapers, but she still fit them in. She was not only less available at home, but also at social functions she spent much of her time with computer friends, rehashing Internet messages and commenting on new Web sites.

Hal got bored at these affairs and ended up drinking more than he should. Whatever happened to card games, table tennis or dart boards? Hal used to dabble in poetry, but now the machines have eaten up his dabble time. *Sadly, he didn't seem to miss it all that much.*

Hal disliked his job on account of all the detailed work. The only thing he really enjoyed was selling. But his company insisted that he take advanced computer classes and even recommended engineering courses. His time on the computer was growing and now rivaled Gail's. Spending so much time at a keyboard made him restless and anxious. The last time Hal was on an airplane it dawned on him that he was sitting in a metal tube, a machine, an iron lung, with only seven inches of head room. He wondered if pre-industrial people would tolerate sitting inside a stuffy metal chamber for hours at a time.

One beautiful day, Hal threw his hands in the air, jumped up and went for a walk in a nearby park; he struck up a conversation with a woman (another bewildered Hunter) who was jogging along the footpath. He and the woman sat on a park bench talking about nothing but the beautiful landscape and the refreshing view of green grass. Hal wondered where his marriage was going. He and the woman agreed to meet again. This same time period was when Gail began to act out sexually.

The psychologist that Dr. Gary recommended, a Hunter, realized that personality differences were also at play in this marriage. Hal's lack of help with household chores was an ongoing source of frustration for Gail. Whenever it occurred to Gail that some chore needed doing, she asked him to do it, but Hal was usually reading the sports page or was otherwise engrossed. He mumbled, "Fine," but didn't really take on board the content or importance of her request. She asked if he was unwilling to do his share, how could he really love her?

What Gail didn't know, and what Hal was only marginally aware of, was that he had a Hunter tendency *to do things in chunks rather than sequentially*, at different times. When he started something, he felt pushed to complete it without interruption. He simply liked to do things all at one time and not have his efforts interrupted randomly, on the spur of the moment, whenever it occurred to Gail that something needs doing. He believed Gail got easily derailed and tried to do too many different things at one time. She still refers to it as multitasking, but he saw it as an irritating lack of concentration.

Gail reminds him each time a task crossed her mind and got the same result. *He's ignoring me*, she thought. He couldn't care less. Gail pushes for emotional support while accusing Hal of not caring. Hal is hurt and withdrawn, while she alternated between tears and continuing to remind him of things to do, which he perceived to be nagging.

The psychologist gave them a few suggestions: For example, Gail can put Hal's chores in a chunk or psychological box for him. Writing down her sequential Gatherer thoughts much as one makes the family grocery list, she makes a list of chores and sets aside a block of time (a chunk) to accomplish the chores. (Or she can just compliment Hal on his beautiful Hunter personality and ask him to come up with a creative solution for this problem.)

When Gail employed this Hunter strategy, she found that Hal worked harder than ever before because he was relieved to be working within the psychological and neurological box that he has grown accustomed to over his lifetime and perhaps over the past 8,000 years of evolution from the time Gatherer and Hunter societies were established. What else did the psychologist discover? Well, for one thing, Hal disliked his job, because it entailed sequential, detailed work and had many interruptions. Like all extreme Hunters, he was a big-picture guy who hated details, whether they're auto parts or computers. That's why he scoped out Dr. Gary's television studio when he first sat down with his television host. Hal's a visual person and wanted to see the staging and get an overall impression of how the show was organized.

Gail's (Gatherer) machine-like tendencies discouraged both of her parents from getting closer to her emotionally. Her obsession with reading and completing crossword puzzles discouraged potential friends as well. Her promotion backfired. She left her capable and secure role as executive assistant in the auditing department and crossed into the abstract Hunter region of fashion and design.

The psychologist helped them with their personality differences, but also recommended that Gail drastically cut her machine time and get involved with her son's video games. Hal was to cut his television time drastically and invite Gail to play golf or join her in some outdoor activity she enjoys. He said, "Fresh air and grass…fresh air and grass."

One of Hal's major complaints had to do with control. Under stress, which was now becoming chronic, Gail increased her talkativeness and need for control. Before we leave them let's take a closer look at control and talkativeness, since these often pop up as barriers to meaningful relationships between Hunters and Gatherers.

Control freaks and talkers.

Your airplane is rumbling down the Atlanta runway prior to takeoff when the smiling lady in the next seat inquires politely about your emerald ring. It reminds her of a ring her late husband gave her.

Her kids love the ring and her grandkids think it's magical. After discussing her children and grandchildren, she reviews her husband's childhood, business ventures and her travels to more than thirty countries (if your eyelids are getting heavy, it's okay to nod off).

Her one-sided conversation continues until your plane touches down, mercifully, in New York City. Unfortunately, we all experience people who talk too much and use their verbiage to *control* those around them. It could be our neighbors down the street, our bosses or even our spouses. But what makes these people tick and what can we do about it? Operating inside a diving bell of their own making, these individuals are difficult to reach and hard to change.

If control is really the villain, should we rid ourselves of this nasty beast once and for all? No. Control can be positive as well as negative. In fact, experienced psychologists will tell you that a good definition of stress is simply *lack of control.* To lead happy lives, we need a reasonable degree of control over our thoughts, emotions and behaviors. Okay, you say, self-control is one thing, but we wouldn't dare control the behavior of *others,* would we? Yes, we need to. Otherwise, our civilization would disintegrate and we'd have three-year-olds playing patty-cake in the middle of the interstate highway.

But who would want to control the behavior of well-adjusted adults? Only tyrants, like Stalin and Chairman Mao—or radical terrorism, which attempts to demolish our sense of security and control? No. Lots of people want to control others and for many reasons. Freud sat in his corduroy breeches in his Vienna office and analyzed strange dreams. B.F. Skinner ran rats through mazes and taught us about reward and punishment. Many psychological theorists have been fascinated with questions of control.

Psychologist Bruno Bettelheim noted that prisoners in Nazi death camps imitated their guards' controlling behaviors. He called this "Identifying with the Aggressor" and theorized that children who identify with a tyrannical parent may display autocratic behaviors as adults. Many sexual predators come from homes where they were sexually abused and controlling behavior can be learned the same way.

If you think Bettelheim's theory sounds a bit Freudian, I agree. But Freud would probably focus on the concept of the father-figure (and today, the mother-figure, as well). When your controlling male boss gets dreamy-eyed and calls you "baby," he may not be the chauvinist pig you believe him to be or even a Daddy Warbucks. He may just think (subconsciously) that he's your dad. In this situation you have a couple of choices: Stand on your own two *adult* feet or manipulate your boss, play the game and possibly get a promotion.

Psychologist Alfred Adler said power is a prime motivator and control, if used wisely, generates power. B.F. Skinner made it simple: *People control others because it works!* If people are rewarded, like rats in a laboratory maze, they'll do more of whatever works and do it over and over again. Only in people's cases, the payoff is money, prestige and power, not Wisconsin cheddar.

Some motivational seminars are all about motivating your work force by *controlling* their behavior. One attendee paid good bucks for advice, but complained that no matter what he tried, his employees didn't seem to listen. He needed a quick fix and he needed it now! The controller said: "Go home, fire 10 percent of your employees and the rest will listen." Brutal? Uncaring? Yes, all of that and more.

Marriage counselors come face-to-face with control factors every day. In her book *Controlling People: How to Recognize, Understand, and Deal with People Who Try to Control You*, communications specialist Patricia Evans believes controllers start as pretenders, like children pretending a teddy bear is real and projecting their own thoughts and motives into this inanimate object. Later they anchor their pretend or dream person in a real person and resist the opinions expressed by that person. Surrounded by a psychological bubble, they're unaware of their own behaviors and motives. This is a good description of how some controllers appear to us: in a bubble and unable to change.[3]

Some of our electronic game addicts are in a bubble, too. Stimulus-bound and mesmerized, young minds are vulnerable to the machine's control.

But why does a child create a teddy bear friend when real friends are more fun? Yes, real people don't always agree with us and they have their own opinions, but they can be as friendly as a teddy bear and a lot warmer. Maybe it's the same reason that people fall in love with their machines. Even though we usually talk about the stimulating and rewarding graphics found in electronic games, one of the things we relish is the control we think we have over these monsters. Being supposedly more reliable and predictable than human kind, we don't have to worry about seemingly irrational surprises that hurt us. It's the age-old dilemma between trusting and taking a chance of being hurt and disappointed, or pulling back where it's safe albeit limited and lonely.

Our machines won't strike out at us or talk behind our backs. They give us some fun and expect little in return. There is something very reassuring about sitting down at a computer console and ordering it to do all sorts of fun, nifty and ridiculous things for us. *We're in control*, and unlike Patricia Evans' teddy bear, we're not projecting some idealized self onto it. No, we're truly interacting with the machine of our choice.

Sure, computers are heartless creatures, but when we're feeling hurt or rejected, they're always there, on standby waiting for their master to beam them up. So maybe in the future, the lonely, sensitive kid who can't trust others will go to video game avatars instead of teddy bears. Not exactly a step forward, I'd say. And, of course, Gatherers like Gail are already predisposed to machines. We saw that in her pre-computer preoccupation with reading and crossword puzzles. Gail's personality also predisposes her to talking. As a Gatherer, she likes language, reading, spelling and all things linguistic.

Talk, talk...talk.

What about talkers? *Do you talk too much*? Are others avoiding you? Do friends and business associates hide when table seat assignments are made? Television commercials tell us it could be bad breath and that mints will solve the problem, but a more likely suspect is what comes out of your mouth in the way of verbiage (or verbal garbage). Like water surging over a dam, what you say may suffocate and immobilize those around you.

Trial lawyers like to invent syndromes to excuse their clients' behaviors. "Run-Away Syndrome" and "Violent TV Syndrome" come to mind. How about *rapid talkers' syndrome* (RTS)? Speech is essential in an advanced society, but, like control, too much speech can be troublesome. Speech can be used for communication purposes, but it can also be used to confuse others and hide our true motives.

The term "Fast Talker" comes to mind. This term implies that something is not on the up and up. Do those suffering from RTS try to dazzle us with words, much as greenhorn students try to fool their professors by adding puff to their lengthy term papers? Psychotherapists in training learn to "listen with the third ear" and break though defensive verbiage. The FBI and the CIA ignore what people say and focus on body language (nonverbal communication). Maybe monks, who take vows of silence, have found something important after all.

I would argue that people who suffer from RTS lack empathy for others or have such an ingrained habit of talking and controlling that they're unaware of their effect on others. They resemble the chronic e-mailer who must share all his or her latest insights or forwards a "can't miss" joke or dubious urban legend for the enhancement of all.

Can't they tell, while inundating us with mega messages at cocktail parties, that we're just plain bored; that we don't care? When we order yet another drink, rather than listen to more of their litanies, and surrender to alcoholism as the lesser of two evils, they still don't seem to get it. And how the RTS people take in oxygen during their long-winded filibusters will remain forever beyond the grasp of science, much like the common cold.

So what is excessive talking? Is it the number of words spoken in a given period of time? No, there are fast talking people who are fascinating to listen to. Is it the number of words per sentence or the lack of meaningful pauses (paragraphs)? Now we're getting closer. Is it the lack of prioritizing and the lack of a logical thread to hold the thesis together? By all means. And who are these people? Are they trying to sell us something?

If speech is basically the process of thinking out loud, then impulsive, confused speech may reflect an impulsive, confused mind. Are RTS's intellectually challenged? They seem to have a great memory for dates, names and trivia. I can recall parents of an intellectually challenged child who were proud that their child was able to remember a bus driver's polka dot bow tie. But this is a bad sign. It shows that the student is over-focused on detail and can't prioritize. No one else on the school bus noticed the driver's tie. It shows potential problems with concreteness and abstract thinking, as well.

On the other hand, some bright individuals who aren't challenged intellectually are eager to pour out their verbiage without restraint. But their outpouring is situation-based, temporary and not a part of their overall personality makeup.

So how did our friends Gail and Hal turn out? As you know, at first Gail's computer helped her reduce stress. She was better organized and able to accomplish more—or so she thought. Her computer efforts were self-reinforcing right up until the time Uncle Gaderian, the pandemic robot, reached up and pulled her into the machine. Hal figured he had helped push her in the wrong direction, so when the psychologist offered him help, he accepted. With some good advice about personality styles and temptations and courageous hard work on both their parts, they returned to their previous slow, warm and dizzy state of human existence…and lived happily ever after.

How to Reconnect

1. Except for necessary business work, "recreational" and other personal use of computers and television viewing should be limited to a total of two or three hours per day (adults). You're missing out on the real world.
2. Never sign your e-mail messages with an ASAP request. It's rude. If it's that important, make a phone call.
3. Respond to "urgent" e-mail requests (if it was really important, the person should have called or used an envelope and stamp) with this standard line: "I have a lazy brain and need to sleep on all requests for information."
4. If you can't imagine the recipient of your e-mail message being pleased with your message, make a phone call.
5. If you're feeling irritation while writing an e-mail, put your message in the *Mail Waiting To Be Sent Box* and sleep on it (or pick up the phone).

6. Volunteer at a school or at a charity. Get out of the machine rut and meet some warm-blooded humans. Kids with problems can lift your heart.
7. If, because of gender or personality, your interests differ from those of your spouse or significant other, spend some time alone pursuing your interests. Don't let guilt stop you. Your spouse may also enjoy the freedom to explore other friends and interests.
8. If your spouse is an extreme Gatherer who multitasks, gets side-tracked and talks too much, use agreed upon non-verbal cues to help your partner prioritize and get back on track.
9. If your spouse is an extreme Hunter, remember to put your requests into a *chunk* or psychological box. Also, show your Hunter how to organize through better use of technology.
10. Extreme Hunter and Gatherer personalities should analyze their job requirements against their personality styles. Have work tasks changed? Are these changes creating stress at work and at home?
11. Date your partner at least twice a month without friends or cell phones.
12. Get yearly calendars to track and organize future social events. A twelve-inch by twelve-inch paper (not computer) calendar, with one-inch to two-inch boxes lets both partners keep track of future plans.
13. Stop forwarding "cute," "clever" and "brilliant" messages on politics and life to your friends. If these messages seem irresistible, and you feel you must share them, limit them to one per week (and practice prioritizing).
14. Slow down and smell the green grass. Take one whole day a week for relaxation and family time. Go to church or go to the beach. Just go (or rest in a hammock).

Illustration courtesy of Dan Piraro

9

Contamination

"Even the Mighty Shall Fail"

Trouble at the top.

In October 2006, Patricia Dunn, chairman of the board of the Hewlett-Packard Corporation, was called before the Investigations Subcommittee of the United States House of Representatives Committee on Energy and Commerce to answer questions about her alleged criminal violations of privacy. She was accused of spying on board members in order to discover who was leaking confidential information to the press. According to an article in *The Wall Street Journal*, Dunn's downfall resulted from her clash with a powerful director on her board.[1] By now, you will have little difficulty in assigning Gatherer and Hunter characteristics to these two adversaries.

Patricia Dunn, the chairman, was described by her opponent Tom Perkins as "a stickler for process and procedure." She brought a careful, rules approach to the company. She was formerly a secretary and became head of an investment firm, where she succeeded in the orderly world of index funds." Her conservative approach to risk management takes much of "the guesswork out of investing." She believes updating handbooks to be important and stated, "Our scientific approach to investigating allows for active control of risk and return." She relaxes with her husband in their modest vacation home. Perkins, according to the article, worried that Dunn would pack the board with "blue ribbon" persons having no entrepreneurial genes. Perkins is a venture capitalist who wrote a steamy novel, *Sex and the Single Zillionaire,* and enjoys life aboard his 289-foot yacht. He formerly worked as a lathe operator. He said, "We know that success is rarely a straight line."[2]

Would knowing the personalities I've assigned to them prior to their board battles have made a difference? I believe if it had been expressed clearly and forcefully, along with the brain personalities of other board members, *it could have made a real difference.* Dunn and Perkins would have understood that many of their mannerisms, comments and behaviors were representative of their personalities and brain profiles, rather than as personal insults or attacks. They would have laughed together at some of their irrational reactions and learned to agree to disagree. This would have allowed for a climate of respect between two people who view the world from completely different perspectives.

Another example of a Gatherer and a Hunter mixing it up occurs in the article "Hijacking a Dream" from *The Daily Mail.* Journalist Tom Bower wrote that an American lawyer by the name of Randolph Fields was the brains behind Virgin Airlines. According to Bower, Fields negotiated with Boeing to lease a secondhand 747 and found retired airway crews prepared to work at low salaries. However, according to Bower, Richard Branson assumed credit for Fields' achievements and then brought in his own flair for promotion. He decided they would announce a 30,000,000 pounds advertising campaign. When Fields protested that they didn't have 30,000,000 pounds, Branson allegedly responded, "of course not, but we'll announce it to get the newspaper coverage and won't do anything more."[3] This resulted in a publicity coup that brought Princess Diana of England to the launch of a new air bus.

But Gatherers and Hunters can learn to live together, even if they don't get along. The year 2006 was the thirtieth anniversary of the movie *All the President's Men* starring Robert Redford and Dustin Hoffman as Watergate reporters Bob Woodward and Carl Bernstein. In a *Los Angeles Times* article, director Alan J. Pakula claimed that Bernstein and Woodward disliked each other.

Nevertheless, they got the job done. Pakula states, "Bernstein could be right intuitively—but dangerous left to himself...Woodward cautiously would have to go from one step literally to another. And yet it was Bernstein's daring that was necessary." Director Paul Pakula "wondered if the charming, handsome Redford, then 39, could play someone so different from himself.... Redford would have to 'scrap his charm. It's that square, straight, intense decent quality of Woodward's that works.'"[4] Hunters and Gatherers at it again?

Over the past five years, government lawyers have vigorously attacked government officials and CEOs of major corporations. Some top executives at Enron and other large corporations have claimed they couldn't remember detailed conversations with staff or the small print on contracts. They pleaded that administrative efficiency led them to delegate authority to lower employees who were responsible for any crimes that may have been committed.

In my opinion, once again we have a direct conflict between Hunter and Gatherer personalities. Through law school, most lawyers follow a Gatherer route, which emphasizes the assimilation of large quantities of facts and details, including case studies, intricate legal precedents and the sequential histories of case law.

There is an old maxim that those law students with high grades become judges, while those with low grades become trial lawyers. This fits our discussion of personality theory to a "T," because academic success is tied to Gatherer skills such as reading, writing and research, while attorneys with Hunter talent, such as former presidential candidate John Edwards, rely on emotional and theatrical talents to hawk their stories to juries.

Unlike government attorneys and judges, or even sensation-seeking trial lawyers, entrepreneurial CEOs get to the top because they disregard details, over-delegate, clear their minds of stubborn memories and focus all of their energies on the really, really big picture. They look to the future, never to the past. Even the present is ancient history and not worth a scrap of intellectual energy.

But why are these guys making the big bucks? The answer is that they're hired to be company prophets. That's right. They're not paid to remember what someone said about line 133b on the revised and unaudited balance sheet on July 27 at 2:30 P.M. They don't claim to be saints, but these *are* the guys who want to buy up all the wells in the desert. If they succeed, they're stars; but if they flunk, they go back to Tulsa with a modest retirement package of $100 million.

Not only do some claim they don't remember a conversation at a meeting that they should have skipped, they state they don't even know what was discussed. Anyone can run a business. But who can create a business that recreates itself and reaches for the stars? Can you do that if you're looking over your shoulder at last year's tax returns? The main reason they may not recall details is because when you don't retrieve and consolidate facts, you sure can't remember them. It's hard to remember something that isn't there.

Does that mean that all these high-powered men and women who slipped up on their tax returns are just innocent Hunter types who didn't pay attention? No. The great recession of 2008-2009 may have resulted from inefficient Gatherer regulators, insipid government mortgage programs and creative—but greedy—Hunters.

And what about the juries in these cases? Because of the complexities of white-collar crime, there may be a tendency to select nerdy digital natives with higher academic credentials who may discriminate against Hunters. But if Hunter types are selected from a pool of citizens that is less educated, blue-collar and small business owners, they're going to be more sympathetic to memory problems, even at the boss level, because they've been there themselves.

Difficulty recalling detail.

Does the Gatherer, bureaucratic lawyer get a little extra gratification from bringing down unbelievably wealthy, high-and-mighty CEOs, who got to the top through social contacts, luck and a few good ideas? Most likely.

While terrible crimes were committed that destroyed the lives of thousands of innocent investors, an understanding of the Hunter personality makes it possible to realize why some corporate directors have difficulties with rapid fire interrogations about brief, low-priority conversations. Surely, jurors suspect the worst when enormously successful executives at the top of their games can't recall the details of conversations with subordinates.

However, several recent presidents of the United States, and other celebrities had difficulties with details. Included in this group are a movie actor, a big-time college football player and the former owner of a major league baseball team. All of whom I believe qualify as Hunter.

When Gary hands Hal an accounting statement, Hal looks at the net profit on the profit and loss statement and little else. He refuses to study the balance sheet or other minutiae. If Hal were indicted in a corporate scandal, he would not do well. The prosecutor and the grand jury would have a difficult time believing that a business school graduate who had reached an upper level in the corporate structure could be so ignorant about corporate financial matters and so forgetful of names. They would believe, much as Hal's teachers of old, that he couldn't have arrived at the correct solutions without dotting the i's and crossing the t's. They would conclude, incorrectly, that he was lying.

The Hunter as hunted.

So what does any of this have to do with the IT world? Are we thinking that our latest cell phones should suck some of the brain cells from freewheeling Hunters, pop their creativity bubbles and push them into the Gatherer zone? That way the risk takers would be under the scrutiny of technocrats and their precise legal and accounting procedures. Would this have helped ameliorate the 2008-2009 stock market crash?

Maybe the IT world is partly to blame. Journalist Peggy Noonan, in a *Wall Street Journal* article, wonders if people have felt disconnected and may be looking to return to something that is human, something of which they can be a part. "For a generation we've been tapping on plastic keyboards, entering data into databases, inventing financial instruments that are abstract, complex and unconnected to any seeable reality. Fortunes were made in the ether, almost no one knows how; there's a sense that this was perhaps part of the problem. Workers tapped on keyboards and produced work they cannot see, touch or necessarily admire. They'd like to make the country better, and stronger, in a way that they can see."[5]

If the machine environment is responsible for some of our economic woes then we'd better get ready for more of the same. After the wave of corporate scandals following the Enron collapse, will changes in the membership of company boards of directors' result in more Gatherers? They very well may.

Revolt in the Boardroom: The New Rules of Power in Corporate America supports this prediction. Author Alan Murray studied eleven old guard CEOs and their replacements. He interviewed Walt Shill, director of the consulting firm Accenture: "For many years, strategy was about determining the future and making big, bold moves. Now it's hands on, getting your hands dirty and delivering the results. People aren't looking to be on the cover of business magazines anymore. In many ways, boards are now looking for boring CEOs."[6]

"In the old days you weren't expected (as a board member) even to know how to read an annual report," says Matthew J. Barrett, law professor at the University of Notre Dame. "Now they will need 'fluency' in law, accounting, or other basic languages of their business."[7] Add in the financial crisis of 2008-2009 and, if the past is predictive, there may be greater regulation over banks, brokerage houses and perhaps government spending.

There will be more Gatherers in high-level positions and on boards of directors. The irony here is that regulators don't stop cheaters and they add to the burden of running businesses. And in my opinion, the growing number of cheaters corresponds with the lessening of our human values. We have replaced morality and ethics with machine thinking and legal and scientific solutions.

Does this mean the Gatherers of the world will come to dominate business management? Probably not. Later in this chapter, we'll discuss how it takes both Gatherer linear and computer skills and Hunter creativity to lead companies. But some of the old-time "play it by the seat of the pants" directors can be seen as risks. The big-picture style of ignoring details is not helpful when government lawyers come calling.

We will still need people in sales and the ability of a salesperson to size up a potential client on a first-time telephone call is not something a computer can do. But will these gifted salespeople still retain this ability after their brains have been hammered with peripheral stimulation in early childhood?

Hunter survival skills.

The creative, nonverbal skills of the Hunter have always been important in the sales field. Often, the illusion and perception may be more important than just getting the facts straight. In Santa Clarita, California, home buyers step through the front door of model homes and find an affable man shouting hello from the kitchen and offering treats to the buyers' children, while his wife presses them to try fresh baked cookies. The couple's children volunteer to show the visitors their rooms.

In truth, this cheerful family of four was a group of professional actors paid to show the buyers how life could be living in the house. A Gatherer can be overwhelmed by visual, emotional, nonverbal communications that are not broken down sequentially.

Speaking of using Hunter qualities to close deals, some American business people traveled to Japan expecting a quick, sequential and well organized approach from their hosts. Sure enough, the Japanese welcomed them. But they put off the negotiations. They entertained the Americans lavishly and took them golfing and sightseeing to slow down the process.

The Japanese executives realized that timing is everything. When there was only a day left they pushed forward their proposals under the pressure of time. These are Hunter business skills at their best. Much of it is about timing, rhythm and factors not directly related to the content of business proposals. They won't go away soon and you won't find them in the machine world. How will our future business people tolerate changes in tempo? Accustomed to game-like speed, will we be pacing in circles and trying to multitask instead of partial-task?

Business leaders have an instinctive feel for opportunities. Sometimes these windows of opportunity open and close quickly and a computer is not known for these gut level hunches. A Gatherer who is relying strictly on statistical data and accumulation of sequential facts may miss the big opportunity all together.

Stressed-out executives used to get a break. At lunchtime they might leave the office, sneak off and hide for thirty minutes and perhaps exercise or read a chapter from a novel. This little vacation not only relieved them from the direct pressures of work, but also reminded them that there was another world outside of the computer and rush of office business. Now executives take the office with them to lunch in the form of smart phones. Oh, and since there's really no office anymore, the executive has no protection from outside interference.

Vocations.

Let's take a quick look at vocations. In 2007, Garrison Keillor said on the *Prairie Home Companion* radio show that the kids in Lake Wobegon were afraid of a Chinese Communist invasion and that the Commies would ask them to choose between denying the Lord and eating a bucket of saliva. One kid said he would deny the Lord but keep his fingers crossed. Another boy said he would cross his toes. When that was ruled out, a third boy said he would cross his fingers in his heart. Garrison Keillor said "this is the point that separates accountants from writers."[8] (Gatherers from Hunters?)

If Keillor is referring to creative writers—who are Hunters—I believe he's right. But even writers vary, according to their personality style. Hunters are more likely to use their imagination to write fiction, whereas Gatherers are more likely to write nonfiction. But even within the fiction realm, Gatherers are more likely to write tightly formulated mysteries, while Hunters are more likely to write thrillers and adventure stories.

Hunters have the potential to become great entrepreneurs, but they're often bored with school because they don't like sequential learning. Homework doesn't turn them on either. As a result, they sometimes drop out of college to pursue some creative idea or other. Unfortunately, the Hunters, who are the best coordinated people around, can't be pilots or surgeons because they can't get through the Gatherer academic obstacles.

At the same time, extreme Gatherers, who are high academic testers or perhaps even members of Mensa, may be operating at low-level white collar jobs because of a lack of street smarts. The

Hunters' best education in college probably comes from fraternities and sororities, where they can refine their social skills for use in the world of work.

Some criminals in England—Gatherers, no doubt—chose the scaffold rather than go to America when a Royal Judge gave them that choice. But other risk takers were pleased to come to the New World. Perhaps this explains how the United States came to enjoy a disproportionate number of creative types. When they reached "New" England, the most adventurous of that lot kept going west. They finally wound up in California, which might explain some of the chaotic and eccentric ideas and laws that come from that state. As debtors, they escaped the hangman's noose, but wound up creating the world's largest debtor's prison: California.

It is interesting how the best intentions sometimes result in unintended and destructive consequences. In England, students have to pass a test at age sixteen to determine if they can enter college. Hunters, who are not as adept at academics, are shuttled aside. In the United States, a system of community colleges and other opportunities keeps the Hunter in the mix. Some folks speculate that U.S. productivity is superior to the continental Big Three: Germany, France and Italy, because of the dynamism in this country.

Nobel Laureate in Economics Edmund S. Phelps writes in *The Wall Street Journal*, "The country's dynamism, being slow to change, is not measured by the growth rate over any short or medium length span. The level of dynamism is a matter of how fertile the country is in coming up with innovative ideas, having prospects of profitability, how adept it is at identifying and nourishing the ideas with the best prospects, and how prepared it is in evaluating and trying out new products and methods that are launched into the market."[9]

Some people believe Hunter type high school students should go to vocational school. In one community, a Vocational Technical Institute was set up to allow high school students to develop vocational skills while still attending academic classes. This would keep them motivated and give them a taste of "hands-on" work. But guess what the Gatherer school administrators did on the way to the market? They made it a requirement that one had to have above average grades in order to take the vocational courses.

Even though Hunters and Gatherers are found in both genders, my experience with business management indicates that most stereotypical Hunters who reach top management levels are men. This may be a function of pure prejudice. Women usually excel in reading, language and test-taking. At the present time, they gain entry to the corporate offices because of high college rankings and the expectation that they will be organized, detailed and thorough.

Today, right-hemisphere Hunter women are more likely to be found as presidents and CEOs of small businesses. If Hunter women aren't hired by large companies, how can they demonstrate their creative abilities? This may be the last glass ceiling for women in the work place, but it can be overcome.

Tips for Hunters:

1. You Hunter types can benefit greatly from information technology. IT offers some of the organizational skills you lack. Take the cell phone, for example: It can be your phone, camera, calendar, e-mail source, address book, Internet browser, map/navigational device and much, much more. Hal the Hunter took his family out of town to a football game. He was to meet some old friends who had tickets to the game, but he'd left his contact information at home. He used his phone to go on the Internet for the name and location of the hotel where his friends would be staying, then used the map to find directions to the hotel, then referred to the stadium seating chart to choose which tickets to keep. Keeping a Hunter well organized is a big job—and just as important, it keeps him out of the doghouse, especially if his wife is a Gatherer.

2. Are Gatherer talkers eating up your valuable time? One Hunter executive developed a strategy to curtail visitors from dropping into his office unannounced to pass the time of day. Fancy technological solution? Persuasive lecture? No. He just used a non-verbal communication that did the trick. After removing one of the two chairs in front of his desk, he piled envelopes on the seat of the remaining chair. This obligated visitors to stand and reduced their long-winded speeches.

3. In the past, Hal would invite a business associate to lunch and write the location and time on a piece of paper or the back of correspondence—whatever was handy. Then he'd lose the reminder in the clutter on his desk. Now he sends a luncheon invitation to his associate, Hazel, via e-mail. When Hazel accepts the invitation, the time and date are automatically loaded onto her computer calendar and cell phone, as well as his own, and they both get a timely reminder.

4. Hal's a sales manager and often out of the office. When he receives a phone call asking for his schedule or someone's phone number, he's got it. One of his assistant managers does after hours work and leaves the office at 4 P.M. When Hal gets an e-mail about an important stock offering, he can relay it to his manager.

10

Disconnection in the Workplace

Coping with co-workers.

Relating to people in the workplace, one has to ask oneself: Do we really know these people or are we stuck with superficial stereotypes and arms-length small talk? This chapter examines Hunter and Gatherer tendencies in the workplace and how the machine process may infect our relationships.

EST is a wireless system that lets diners at restaurants page their servers whenever they have a need for necessities like an extra napkin or another drink. The system also lets management check the server's watch-like wristbands to know when meals are ready. Management can track how long it takes the server to wait on customers and how many times he or she has to be buzzed. Next will come a machine that tells management what you eat, how fast you eat and when to clear the table for the next customer. Take your time and enjoy your meal!

More and more retail stores are requiring lengthy, computer scored questionnaires prior to even interviewing job applicants. Hunter applicants may never get a chance to show their sparkling personalities, because the Gatherer's robotic tests slow them down. Our skilled craftsmen can't get jobs at some large home supply stores. Machines love craftsmen almost as much as they love artists and poets (which is not at all). Recently some enterprising Hunters found answer keys to tests and sold them on the Internet. The only way test owners could get these individuals to back off (offline) was to claim copyright violations. When values go, lawyers take control.

But we may not need to worry about questionnaires to qualify job applicants in the future. According to an episode of *60 Minutes*, researchers are working on an electronic BEAM that can read thoughts. This beam would replace the need to cram people into MRI specimen machines to examine them. Researchers are even able to detect feelings of lust in their volunteers and can tell exactly what shoppers are thinking.[1]

According to an article by Clive Thompson in *The New York Times*, surveys have shown that many people would rather deal with ATM machines than bank tellers. Some Gatherers would rather go to a Web site than deal with a living, breathing employee. Of course, today the living and breathing employee is attached to a computer and he or she often talks like a computer. And, they are busy. A survey of workplaces in the United States found that workers were interrupted and distracted roughly every three minutes and that people working on computers had an average of eight windows open at one time.[2]

But employees are expensive and only want to work forty hours a week. They also require sleep, vacations, benefits and other insufferable demands, so we're replacing them with machines. Sometimes we have a choice between a machine and a Web site. Maybe those icons of smiling faces on Web sites are put there to make us think warm and cozy thoughts, but the research shows we're better off with face-to-face relationships.

Computer love.

At the office, Gary, the Gatherer, loves his computer. In fact, he is his computer. If Gary is asked to set up a meeting of five busy persons who are hard to pin down, he simply goes to his computer and sends them e-mails letting them know there will be a meeting and the time and place.

When his boss, Hal, the Hunter, asks him a day later if the meeting is set, Gary is happy to report that he did as his boss requested. Hal is surprised that Gary accomplished this difficult job so easily. When Hal asks him who is coming to the meeting, Gary admits he doesn't know. Actually, Gary did nothing but write e-mails. Yet, he is proud of his efforts. He suffers from computer attachment syndrome.

Gary doesn't know about a new system that tracks workers' business calendars. It sniffs out the open times in their schedule and tracks each person's calendar. Colleagues are "invited" to attend meetings when the system lets the boss knows they're free. Of course, this also snuffs out the workers' personal space. No having coffee with the new truck driver or sneaking off to buy little Heather an ice cream cone. Not anymore. Not when Big Brother has electronic eyeballs.

Gatherer over-reliance on machines shows up one day at a beauty parlor. Hazel, a Hunter, is having her nails polished when she hears Gail, a Gatherer, a young adult a few chairs away, ask if daylight saving time would start the following week. Hazel and the hair dressers laugh, because the change to daylight saving time had taken place the week before. Gail is surprised. Relying entirely on her cell phone, she has no clocks at home and never wears a watch. Her cell phone had automatically adjusted for the change in time, and she didn't notice she'd lost an hour of sleep or that it was dark outside when she awakened in the morning. This is a far cry from Hal, the extreme Hunter, who eats only when he's hungry and adjusts the grandfather clock by looking at the moon outside.

At the same time, computers are godsends for those Hunters like Hal. To paraphrase Winston Churchill, never have such tiny, sequential machines done so much for so many big-idea people. Without the need for neurosurgery, the computer provides Hunters with the Gatherer-like personality they never had. Oprah Winfrey, who displays Hunter characteristics, sometimes uses machines on her television show to reinforce emotional connections.

Like his beloved computer, Gary speaks in a dry manner, with little expression or rhythm. He doesn't adjust his voice level to fit the circumstances. He doesn't have an "office voice" or a "tennis court voice" or a "home-with-the-family voice." Because of his robotic verbalizations, people who call occasionally hang up, because they actually think his voice is an answering machine. He's much better than Hal at remembering names, but he has more problems remembering faces and spatial directions.

Gail often feels excluded in the workplace and doesn't know why. She works hard and is always polite to others, but sometimes her communications are misunderstood. Gail's problem is with nonverbal communications. She's a good writer and can make her computer hum, but, like her computer, she's in the dark when it comes to sending and receiving non-verbal messages.

Gary misreads Hal when Hal, highly enthusiastic and beaming with pride, comes up with a creative idea. Gary thinks this guy is off on a tangent and this could cost the company some big bucks. Gary earned his position through sacrifice and hard work. He learned to analyze facts and use logic to arrive at realistic conclusions. Now here's this guy Hal, who didn't do all that well in school, and he's trying to give Gary and other staffers fanciful concepts that may have sex appeal, but probably won't work.

This seemingly effortless ability to create new ideas often brings praise from supervisors and co-workers alike, but it can generate envy and jealousy among colleagues. Gary wonders why Hal should be rewarded for sitting in his office staring into space while he, Gary, is grinding it out, day after day, and following company rules to a "T." Poor Gary must feel like the computer that lost to Kasperof.

Hal does more than stare into space; sometimes he goes to online gambling sites and has to check frequently on his fantasy football team. But much of his creativity does pop out in sudden bursts or epiphanies. This makes Gary uncomfortable. *But let's go easy on Hal.* He even works when he sleeps, because some of his best (and worst) ideas may come to him when he's asleep or semi-conscious. How do you teach *that* in business school?

Gail, an extreme Gatherer, will never be accused of fraternity house antics. In fact, she may adhere too closely to office protocol, even when bending the rules would make sense and help the company. A customer enters a take-out pizza parlor early one rainy night, when there are no other customers. Gail takes the customer's order and then announces in a loud voice that it will take about three minutes and that his number is fifteen. Gail gives the customer a slip with the number fifteen printed on it and points to a table that displays a "Take-Out Customers Only" sign. She tells the customer to sit over there until the order is processed.

No other customers enter the pizza parlor during those three minutes, but when the pizza is ready, Gail shouts "Number fifteen!" in a loud voice and holds out her hand for the numbered slip. Don't ask Gail for anything unless you're willing to go through channels and fill out the proper forms. Five copies, please.

When Hal asks Gary a pointed question, Gary has a difficult time responding in a succinct manner. He tends to respond with already formed little speeches that are tucked away in his computer-like memory bank. Gary's response includes the question presented, but it also contains extraneous information that is not on the precise track Hal is operating on at that moment. Their respective styles of communication are so different, it's a wonder they can communicate at all.

It's not about me. It's about us.

So far, we've talked about control of individuals, not groups. Control factors for office and factory workers in the U.S. are similar to those for workers in most other countries, even Communist countries. The goal is to control individual motivation and effort for the *good of the group*. Bobby Bowden, coach of the Florida State University football team, used the following slogan in 1999 when his team went undefeated and won the national title: "It's not about me, it's about us." He used it again for the 2009 season.[3]

Hal is willing to tolerate team meetings as long as they are brief and to the point. But he draws the line when it comes to group brainstorming. He believes the team approach to creativity is distracting and non-productive. Hal may be right. Psychologist Paul B. Paulus, Professor of Psychology at the University of Texas, found that "there are so many things people do in management because they think it's good, but there's no evidence for it. Teamwork is one example. Brainstorming is another."[4] He conducted a study comparing brainstorming sessions of four people who brainstormed together versus brainstorming separately. Typically, group brainstormers perform at about half the level they would if they brainstormed alone.

Hazel is not into control. Her ability to prioritize based on a worldview also makes her less reliable and predictable as an employee. Remember: She generates her own rules. Don't be surprised if she takes a jaunt to India to visit a Buddhist monk or flies air balloons around the world. Unlike Gail, the Gatherer, some extreme Hunters participate in frat house antics because they're creative and playful and need to relieve boredom.

Sometimes their high production rates protect them from censure, but they can raise the ire of fellow employees. As Jared Sandberg reports in a 2006 article from *The Wall Street Journal*, these cut-ups survive by the following rule: "Anyone bringing home the bacon can behave any way they wanted."[5] When management does criticize them, they respond with "You want me to close deals or play nice with the crybabies?" "Well, both," the manager responds, "but I'll take the closed deals first, thank you." High producers will always do well, but what if the growing emphasis on Gatherer mentalities keeps Hunters out of the workforce?

One company, recognizing the increased rigidity and mechanization of the office environment, encourages "productive regression." For a meeting requiring creative thinking, a toy maker built a meeting room to resemble a tree house, complete with a large artificial tree trunk sprouting through the floor. According to an article in *The Wall Street Journal*, "The tree-house room 'was designed as a place that brings out the boy and girl in us,' says John Duprey, Mattel's director of organization development, who says the design helps employees think unconventionally."[6] If the mechanization culture continues to grow, we will see an increase in encounter groups, primal therapy, Rolfing and other quick fixes to lubricate our robots and free our souls.

Gary's need for precise numbers slows down his production but increases his certitude and reliability. Hal, on the other hand, lives a life of approximations, guesses and hunches. Gary and Hal attend a seminar where the speaker reports on a revolutionary new procedure that dramatically increases the production of mattress covers.

"We used to average six per hour," the speaker says. "Would anyone like to guess the new production number?"

Hal responds immediately. "Twenty."

The speaker laughs and nods his head. Twenty is the correct number. Gary is astonished. This is magic!

How did Hal pull this number out of thin air? Even Gary's trusty computer wouldn't know where to start. *Hal's not actually sure.* But if he broke it down sequentially, which he's not inclined to do, he might have figured that doubling production would not *match the speaker's level of excitement and enthusiasm.* How about tripling production and then adding a couple for good measure? Hal uses non-verbal cues to produce a near-spontaneous response and isn't afraid to take "educated" guesses. Can a computer do that?

The music of early Russian nationalism broke away from German and French romanticism and was characterized by its spontaneity. It was stated that the Russian composer starts with the conclusion and *just might tolerate logic,* whereas the German composer had the music all laid out and systematized prior to composing the symphony. The Russians stated that "the German chews over what has been said, but we are not cud chewers."[7]

Hal likes to paraphrase a story he heard about the Buddhist tradition. Gary, a Gatherer student monk, told the Buddha that he would not become his student until the Buddha answered his questions about whether the soul and the universe are eternal. The Buddha responded: "I am offering to take a poison arrow from your arm, and you are asking me what type of poison is on the arrow and what kind of feathers guide its flight, and what the arrow's shaft is made from? Let us put these things aside and move forward."

When Hazel is in a group meeting, she is most creative when permitted to suspend belief in what is being discussed. Hazel feels uncomfortable when she's pushed into the trees and can't see the forest. She consciously and unconsciously pulls back from logical, sequential thought, which she sees as a trap. Hazel needs to be allowed to ask herself "What if?" questions. What if the boss is wrong about that? What if we turned it on its head or came from the exact opposite direction? What if this solution helps now, but not down the road? What if we shouldn't even be discussing this in this way? What if we dropped this product altogether?

Next, let's look at a paraphrase of a television commercial that demonstrates the result of this type of questioning and creativity. The boss sits at a table with a dozen executives and laments about expenses and how to cut them. All current costs are laid out on reams of paper and copious charts.

The Gatherer boss is desperate. He wants solutions and he wants them now.

The Hunter of the group asks, "What about this?" and points to the desk.

The boss looks at Hazel Hunter as though she's crazy. "What about what?"

"This stuff," she replies, "all this stuff, all this paperwork."

"What about it?" The boss, Gatherer Gruff, is growing irritated.

"It takes time, effort and people. It's costing us millions," Hazel replies.

Money grows on trees.

Ordinarily, Hal is not driven by quick financial gain. He may be full of grand ideas and expansive thoughts because he derives his major pleasure from the act of creativity itself. While money is a very good thing and he would like lots of it, he may not be willing to devote the necessary time to make it happen, especially if it curtails his time for creative thought. His ability to prioritize based on a worldview may also push money down the list as an important priority. He could even be a little sloppy with his personal finances and be overdrawn on his checking account from time to time. His Gatherer friends just shake their heads when Hal makes questionable loans that may or may not be repaid. For Hal, money does indeed grow on trees. He shakes down the creative insights and down come the shiny apples!

The future.

Of course, machines do speed up work at the office. Right? A management type told me that when he walks down the hallway at work he invariably finds several people busy on their computers, but when he looks at what they're doing, they're mostly on the Internet and they're not doing company work. Whatever savings might accrue from the use of the computer may be offset by the temptation to indulge in immediate gratification that carries over from game usage.

Both Hunters and Gatherers who are able to master advanced video games and not abuse them may be learning some problem-solving skills, but we still don't know to what extent these skills can be taught or how they generalize to the business world. The learning density or *amount of time needed* to learn these skills is not impressive, if we are to believe The Federation of American Scientists referenced earlier, but perhaps some have benefited. Many employers continue to complain about poor verbal and numerical competencies from both high school and college graduates. Even teacher applicants send in letters and resumes filled with spelling, writing and grammatical errors.

Perhaps the more intelligent students have picked up sufficient knowledge of basic skills and are benefiting from problem-solving tasks in complex video games while less skilled workers are skipping over the basics in part because of IT distraction. As indicated earlier, you can't conduct the symphony orchestra until you learn how to play the fiddle.

Some of the excitement over cyber-native sophistication may not pan out as advertised. Our Gatherer whiz kids should find themselves right at home in libraries, where protecting material in digital form is an ongoing process. Right? In the *New York Times* article "Digital Archivists, Now in Demand," Victoria McCargar, preservation consultant at U.C.L.A., discusses the merging of library science, preservation and technological ability. "'People with I.T. backgrounds tend to be wrong for the job,' she said. 'They tend to focus on storage solutions: "We'll just throw in another 10 terabytes on that server."'"[8] The result is that wax builds up and buries useless files that make it harder to find the needed files.

No doubt IT is affecting the workplace in both positive and negative ways. Ron Alsop's article, "The 'Trophy Kids' Go to Work," reports that "Millennials," people born between 1980 and 2001, want to shake up their boring workday by listening to music and leaving their cell phones on. They like clearly defined rules, step-by-step instructions and immediate rewards.[9] Maybe they think work is an electronic game. We wouldn't want them to be bored…would we?

Ten Tips to Help Gatherers at Work:

1. Practice prioritizing. Make a game of selecting the top three priorities from a list of a dozen similar goals. Then select three items, in order of least relevance, to drop from the list.

2. Learn to consolidate. Take a one page synopsis of the company's history and goals and boil it down to one paragraph, with a maximum of three sentences.

3. Take a course and learn to speak on your feet.

4. Master relaxation. Get a muscle relaxation tape and imagine the boss walking toward you. Breathe deeply; release tension.

5. Resist the impulse to tuck in your shirt and tighten your belt. You're a person, not a machine. Let it all hang out…a bit.

6. Warm up your office space with plants (green ones), flowers and pictures.

7. Ask yourself "what if" questions about *everything*.

8. Raise a creative thought at a meeting. Start at a small, relatively safe meeting and preface your "wrinkle" with this statement: "I'm just brainstorming here, and this may be way off base, but, ah, what if…"

9. Stay on track. If a comment reminds you of an interesting story in another subject area, question whether it really pertains to the bottom line. If it doesn't, drop it…pronto.

10. Don't let your computer swallow your lunch. Yes, you feel most comfortable and safe hiding inside its steely skirts, but to grow, you need to seek out new attires, even if they look intimidating. We know you like to attract attention, but take down the embarrassing videos from the annual office party.

Ten Tips to Help Hunters at Work:

1. Think of other people's feelings and tastes. Don't put up an offensive picture or saying just to attract attention.

2. Start reading signs and peruse details in notes and invitations. Just following the big picture has not gotten you where you want to go. So now you need to see the smaller particulars and try to understand diverse points of view.

3. Scatter things around your office with joy and abandon, but twice a day, at noon and 5:00, straighten up your workspace.

4. Don't take rules personally. They help maintain structure so try to follow most of them.

5. When the extreme Gatherer irritates you, remember two things: Bullying is out, and you need the Gatherer in order to succeed.

6. Be on time. Making others wait shows a lack of self-discipline. Resist the impulse to slow down when it appears you might arrive on schedule. Arriving at the appointed time shows respect and seriousness.

7. Ask your Gatherer acquaintances to help you sharpen your technology skills. You need them to help you organize and to follow up when you over-delegate.

8. Don't exaggerate. Sure, people like your stories, but the more they laugh, the less they may trust you.

9. When long, detailed meetings drive you crazy, take out your pen and scribble—but not directly on the table.

10. Sometimes your Gatherer bosses and fellow employees miss the forest because of their preoccupation with the trees, but as a Hunter, it's your job to help them see the whole terrain.

Illustration courtesy of Rebecca Skelles

11

Gender Catharsis

Digital Styles in Men and Women

Gender differences seem important to any true understanding of the internet technology phenomenon. Daniel Ruth, columnist for *The Tampa Tribune*, wrote in a December 2000 article about new "research" which revealed that men listen with only half their brains. He went on to say that "women must use both sides of their brains to grasp conversations and the poor things still can't understand why *The Terminator* is such a great movie and why it is important to watch three football games in a row on Sunday." He concluded that the only two essential words for men who are yearning for domestic tranquility are "Yes, Dear."[1]

A recent hardware company advertisement showed two men in a bookstore reading self-help books with titles such as *Healing Your Soul with Aromatherapy*. The caption read: "CUT THE CRAP. THERAPY IS A PUSH-BUTTON PLIERS AND SOMETHING BROKE TO FIX. 'Hey, Freud, analyze this: The plier jaws of this stainless-steel breakthrough snap open like a switchblade.'"

Why do many men have an addiction to football and hunting? Cap'n Charlie Croaker, in Tom Wolfe's novel *A Man in Full*, "loved the way his mighty chest rose and fell beneath his khaki shirt and imagined that everyone in the hunting party noticed how powerfully built he was."[2] Is it the addiction to a romanticized self-image that makes males feel more manly?

These are, of course, stereotypes of the male condition, but there are plenty of prejudgments about women, as well. As Professor Higgins complained in the musical *My Fair Lady*, "Why can't a woman be more like a man?" Women are held to be more social, more emotional and more nurturing than men. And men think women talk too much. But women can be brief when the situation calls for it.

The older generations had the habit of making quick phone calls because of the relatively high cost of long-distance calling at that time. Most women made calls that sounded something like this: "Hello, dear, this is Mom. We'll be arriving on the noon train from Boston. Wear something nice and we'll have lunch before touring the campus. Bye-bye." So response differences may depend on the task at hand as well as the individual's Gatherer or Hunter personality or gender alone.

Most women will emphasize sharing when the situation calls for it, but they can switch to problem-solving when indicated. Perhaps many women seen in private counseling are a selective population of extreme Gatherer women married to extreme Hunter men.

Males aren't the only ones who find it difficult to express emotions verbally or through touch. Some women have the same problem and some women (like some men) truly don't know what they want. Clint Eastwood's movie, *The Bridges of Madison County,* prompted a surge of traffic to marital counseling offices. Women were looking for tall, handsome guys to ride into town and sweep them off their feet.

Counselors called this the "Bridges Syndrome"—this is akin to the "Pretty Father Syndrome"—females looking for someone as strong and supportive as their idealized father figures, who will enkindle their souls with gentle understanding. It might have been cheaper to just buy a ticket to Hollywood.

Indeed, much has been written about gender differences in communications and how they affect the marital state. John Gray, Ph.D., in his book *Men Are from Mars, Women Are from Venus,* highlights significant gender differences in communication. According to Gray, men listen to gather information in order to solve problems, while women listen to relate and share. He believes these differences are genetic and could go back to experiences in antiquity.[3]

Gray's theory could be correct. In Hunter-Gatherer societies where modern humans developed, there may have been a clear distinction between genders in terms of their daily survival skills, with women filling the role of Gatherer and men filling the role of Hunter. As cave dwellers, women were child bearers and nest defenders. They needed to notice changes in small details and develop verbal skills to protect themselves from physically stronger males. Male hunters developed spatial ability and large motor coordination. They had to make lightning quick—what we would today call impulsive—decisions in order to kill for food and feed their families.

Is this why the incidence of hyperactive and impulsive behavior is so much higher in males? To what degree differing sex roles in early human societies continue to shape personality style is impossible to know. Some evolutionary psychologists suggest this could be the case.[4] But other psychologists believe this is a dubious premise.[5] Maybe we're still working through these ancient, horrific experiences. In soap operas, for example, there is duplicitous manipulation and emotional confrontation where gorgeous cardboard characters float on cotton candy clouds of improbability and verbally confront each others' lies and deceits.

Are some women really addicted to these melodramatic productions? If we draw on analyst Carl Jung's theories, could this be much-needed therapy for the collective unconscious, the storehouse of latent memory traces inherited from our ancestral past? It seems logical that these early experiences would have shaped the architecture of the brain.

These ideas about hemispheric personalities and gender differences call into question the stereotypes about women found in books written by counselors in clinical practice who imply that *all* women insist on emotional sharing at the expense of problem solving. Critics of HBO's *Sex and the City* claimed it was written by men, but the executive producer, Michael Patrick King, thought this perception was fueled by sexual politics. "I think people... never heard women speak like that (before)," King said. "It's so interesting that in order for women to be confusing and interesting and strong and sexually aware, they [the critics] had to make them written by men."[6]

This is the way prejudice (prejudgment) begins. Identify the extremes from each of two groups as representatives of their classes, e.g., most blacks are good at sports and most whites have high SAT scores. Creative people are "spacey." All Gatherers are responsible souls while all Hunters are unreliable slobs.

Instead of such generalities, let's focus on a few "solid" differences between the genders that *might* influence their interaction with the IT learning environment: Certainly, some of the following differences between the genders are valid, although there is no clear or sustaining evidence that these inferences are totally correct—as is usually the case with modern science.

Love and sex.

Most researchers believe that women are more social and nurturing than men and that men are more aggressive, both socially and sexually. The radio talk show host Dennis Prager summed it up this way: "Men love women, and women love *a man*."[7] (my italics)

Most men are inclined to keep their emotions under wraps. Remember our friend Hal? He held on to an ingrained belief that disclosure of emotions would weaken his macho assertiveness and toughness (and make him vulnerable). Maybe he has a point. Many people were disappointed in April 2007 when a British Commando admitted that he sobbed each night, because his Iranian captor teased him.[8] This accusation unnerved him.

Psychologists tested groups of women and men for their ability to recall highly evocative photographs three weeks after first seeing them, and found that woman's recollections were 10 percent to 15 percent more accurate. "The wiring of the emotional experience... into memory is much more tightly integrated in women than in men."[9]

In her book *The Female Brain*, Louann Brizendine, M.D., asserts that females have more brain circuits for reading emotions and social nuance as well as for nurturing, communication and using both sides of the brain.[10] Other experts have speculated that women look at faces to build social bridges and for protection, whereas when men perceive faces they see only competitors. But at least one study, according to University of Hertfordshire Professor Richard Wiseman, showed that women were not any better at reading faces than men, although they thought they were. 77 percent of the women said they were highly intuitive, compared to only 50 percent of the men. But women's intuitive judgments were not better than men's. Women identified a real smile correctly in 71 percent of cases, whereas his men did so in 72 percent of them.[11]

As we mentioned earlier, such findings are somewhat contradictory. Such women aren't any better than men when it comes to *identifying* evocative expressions, but they may be better at *remembering* them? Maybe these emotional memories mean more to women than to men. We hope so.

In the 1970s and '80s, some researchers speculated that male aggressiveness was not an inborn trait but developed from early environmental factors.

At least one study has found no differences between men and women in terms of overall *quantity* of speech. As reported by *Science Magazine*, Psychologist Matthias Mehl had 396 college students wear microphones for two to ten days.[12] The recordings showed no statistically significant differences between males and females, but of course the college population is a select group. It may be that the verbal ability and the intelligence required for college work created a selective factor where there were more Gatherers than Hunters and perhaps this overrode any gender differences.

Men focus on small details and show more emotional detachment. They are also more responsive to sexual fantasy and think about sex much more frequently than women do. According to a *Pediatrics* article by Tori DeAngelis, 38 percent of sixteen- and seventeen-year-old males visit porn sites intentionally versus 8 percent of teen girls.[13] Women's sexual fantasies often involve love and commitment. This may account for the popularity of soap operas and romance novels. Men tend to visualize direct sexual encounters, while women embrace intuition and emotional factors.

Some of the complex video games that are *without* sexual content activate areas of the brain that are also activated by sexual images (think about that). Video games that home in on the sexual areas of the brain could lead to the *instant* addiction of the teenage boys. No wonder they're spending so much money on electronic games.

Anthropologist Mizuko Ito's research indicates that girls gossip more while using IT for social reasons.[14] While the girls are the agents for gossip, they claim that boys create more of the gossip. Of course, girls like boys to leave messages on their Facebook "walls". This could leave girls open to manipulation by boys. In the real world, that manipulation is a little more difficult. It might require face-to-face interaction, arranging to meet, or taking a girl to dinner or the movies. Fathers are much more active in coaching and playing with their children's electronic and video games. Hardcore, sophisticated gaming is dominated by boys and men.

S.R. Stern's study of gender differences online shows that girls communicate personal information, feelings and romantic relationships on their Web pages, while boys reference friends and family more often. Boys prefer fantasy/violence, sports and action/adventure computer games while girls prefer educational, action/adventure and entertainment-type games.[15] Women are more likely to e-mail friends and family to share concerns, forward jokes or plan events. They use the Internet to enrich their relationships. They are more likely to download online map directions while men look for news and financial updates as well as sports and video games.[16]

Gatherer—Hunter: left brain, right brain.

Women draw more from both sides of their brains. Based on some research and observation, girls definitely have a more balanced and stable use of both left-hemisphere (Gatherer) and right-hemisphere (Hunter) strategies.[17] Men are more likely to represent Hunter and Gatherer extremes, rather than staying toward the middle of that continuum. This is no giant surprise to most women, and might explain why men are healthier and live longer when they are happily married. Women provide the balance and security men seem to need.

In the old feature film-turned-sitcom *The Odd Couple*, Felix Unger was described as the "fussy" photographer and Oscar Madison was depicted as the "sloppy" sportswriter. But these men really portrayed extreme Gather and extreme Hunter personalities.

Was it their differing personalities that made them so interesting, lovable and "odd"? No doubt.

Anecdotal evidence tells us that left-handed people are more creative than right-handers. Musicians, painters and writers are significantly more likely to be left-handed than those in control groups. We usually associate the right side of the brain with creativity. Recent studies show that strongly symmetrical brains, where functions are not localized to one of the hemispheres, are the most creative. While it may not be as efficient to use both hemispheres, because the information has to shuttle back and forth between the hemispheres and is more prone to errors, more novel solutions are likely to be encountered along the way.[18]

In the *Psychological Bulletin*, an analysis of 144 handedness and laterality studies accounting for a total of nearly 1.8 million people illustrated that males are about 2 percent more likely to be left-handed than females. If 10 percent of the female population were left-handed, then just about 12 percent of men would be as well.[19]

Spaced out.

One of the best substantiated differences between genders is found in what is called spatial ability. This is the capacity to visualize, picture, shape and position, geography and proportion accurately in the "mind's eye." It's imperative for three-dimensional objects or drawings. This might explain why males dominate the game of chess, even in countries where the game is a national sport played by both sexes.

In a study of male and female teachers, according to Ann Moir and Jessell Davis, authors of the book *Brain Sex: The Real Difference Between Men and Women*, twice as many male teachers teach information technology, sciences and chemistry as do women, and the contrast in physics is 82 percent men and 18 percent women.[20]

Meanwhile, girls learn to speak earlier than boys and are more fluent in preschool years. They read earlier and do better with grammar, punctuation and spelling. Later, women find it easier to master foreign languages, while stuttering and other speech defects occur almost exclusively among boys. While women may not be able to pick out an authentic smile any better than men, they're better at picking up social cues when one includes tone of voice and intensity of facial expression.[21]

Looking at cognitive styles on the Myers-Briggs type indicator, significantly more men than women prefer objective evaluation of experience. In one random sample of 659 women from 2,000 households in 300 counties across the United States, only 34 percent of the participants were thinking types, not feeling types. In another large sample, 32 percent were thinking types. In a sample of 6,814 male college graduates, 70 percent were thinking types.[22]

What about developmental differences? Current research shows that boys and girls show processing differences during and after elementary school. In *American Psychologist*, cognitive psychologists Elizabeth Spelke and Ariel Grace say, on average, girls and women excel on tests of *verbal* fluency, *arithmetic* calculation and memory for spatial *locations* of objects, while boys and men excel on tests of verbal *analogies*, *mathematical* word problems and memory for *geometric configurations*.[23]

Gender and IT.

It is important to realize that these studies rely on averages. This makes it difficult to predict what a single individual will do in a given situation. Let's see if we can apply some of this research, along with our Hunter-Gatherer personality theory, to our lives with machines.

As we have seen, social sites like MySpace and Facebook are more popular with females. Female Gatherers should demonstrate the greatest usage. They are disposed to machines and machine interactions and enjoy accumulating the names of dozens of "friends." If they are depressed or lonely, these sites can help them feel more a part of things, but don't necessarily result in new friendships.

Female Hunters, who are more outgoing and action oriented, will use the sites selectively to keep track of real life friends and to foster new relationships. They could care less about the gadgetry. They'll also exaggerate their attributes in pictures and written postings. Hunters might risk the use of texting during class and in other prohibited situations.

Males will use these social sites to a lesser extent, but male Gatherers will use them more than Hunters, and Hunters are more exploitative in their social site and texting contacts. Maleness overrides personality differences when it comes to aggressive/fantasy video games. Males just like them. Male Gatherers are perhaps more fascinated with the mechanical aspects of computer games, cell phones, computers and printers. And the more gadgets, the better. Game content, whether watching, playing or sending messages, is less important to the male Gatherer. This is also true for the female Gatherer, although for females, the gender preference for socialization can override the personality preference for gadgets.

Male Hunters are not tempted by mechanization itself. They can't wait to get into the action games where they can hunt and kill at will. Gatherers seem content to follow game rules while Hunters are more likely to challenge the rules and attempt to rewrite programs and introduce unique solutions to more complex games. They may not have the persistence to stick it out, however, and may give up when the game becomes too repetitive and predictable. They would prefer to start at the game's highest levels, without going through the *sequential steps* to get there.

Female Gatherers are more into cell phones than female Hunters. Male Gatherers talk as much as females, but the communications are more practical and less social and emotional. Male Hunters will use cell phones as a means to an end. To them, outcome is more important than process.

Even though genders differ—sometimes in mysterious and inexplicable ways—studying them in the context of Gatherer and Hunter personality types sheds some light on where we're going. Perhaps both genders can agree to disagree.

12

Physical Therapy

"We Are What We Do."

There's a *Doonesbury* cartoon by Garry Trudeau that features a depressed adolescent who claims she's feeling fine, but sits, bent over, holding her arms tightly. When her therapist asks if she's ever heard of body language, the adolescent responds: "Can we leave my body out of this?"[1]

Unfortunately we can't leave the body out of it, even if our friendly computer would like us to. No, the mind and the body are forever entwined. An acquaintance of mine studied at Cambridge University in England. After two years of study, students were required to take a two-day test that would provide their only grade and determine their future opportunities.

The night before the big exam, they dreamt of various successful scenarios, including carrying their smart suede briefcases to Number Four Downing Street: having a two martini business lunch in London while checking on Greek freighters in the Mediterranean; attending Ascot and cricket finals against South Africa. But if they didn't do well, nightmares included working as a civil servant, hotel concierge or a bank clerk.

How did these young students prepare for this critical evaluation? They started physical workouts months in advance of the exam. To sit through this grueling two-day exam, they knew they needed to prepare their minds and bodies. They lifted weights, rowed up the river Cam and followed exacting diets.

In fact, the idea for this book came from my observations of people in motion; not what they were thinking, but what they were doing. I was sitting at an outdoor café in Budapest chatting with a friend when he asked me to tell him about the people passing by. I thought he was joking, at first, but began to describe my impressions of each person based on their gait, dress and general bearing. I didn't have the opportunity to follow up on my "mind-reading" act, but later, on a cruise ship, I made similar observations and later was able to learn about their lives. It is possible to learn a great deal about people by simply observing them in motion.

Former FBI agent Joe Navarro specializes in nonverbal communications. Pulling on one's collar shows concern, touching the fingertips of both hands shows confidence and inter-lapping the fingers shows anxiousness. Mr. Navarro makes an important point about nonverbal communications and our need to learn them. As mentioned in the article "Signal flair," published in the *St. Petersburg Times*, he says, "I wanted to get people to observe the world around them, to enrich their lives, to be able look at their children and say, this child needs a hug now, this child has something he wants to talk about or that one's having a tough day. That's what I wanted to see."[2]

In fact, the Fox Network currently has a show called *Lie to Me*. It's based on psychologist Paul Eckman's studies of facial gestures and expressions. One of the people Eckman studied was superstar baseball player Alex Rodriguez in a television interview. Eckman noticed a number of gestural slips, like raising one shoulder or half-shrugs, which gave suspicions when he denied using performance enhancing drugs.[3]

Ginger, a sixteen-year-old Gatherer, wants what most sixteen-year-olds girls want. She's searching for a boy who is cute, likable and trustworthy. One day, when she's at her school locker, she a sees a boy with an impish grin, looking her way. His face "seems to talk." Now Ginger is presented with a choice. She can evaluate this young man by texting and checking MySpace to find out more facts about him, or she can *really* look at his expression.

If she has practiced and is able to size people up based on their movements and expressions, she may be able to tell in a couple of milliseconds (computers can't do it at all) whether this smiling boy is sincere or whether he uses his *public* self to manipulate pretty girls. If she entered the IT world at age four or five and has spent thousands of hours on her computer and other tech devices, she may not be practiced at evaluating faces. The machine has rewarded her for keeping her eyes on the periphery. She could even walk past him without seeing his face. Or, because she is so distracted with her multitasking, she might not see him at all. That would be a shame!

Ginger needs to read Paul Ekman's book *Emotions Revealed: Recognizing Faces and Feelings to Improve Communication and Emotional Life.*[4] Because she's been exposed to the IT world, where she needs constant reward and stimulation, Ginger might find the book a bit on the boring side, but her friend Heather, a Hunter, would enjoy the visual format.

Ginger and Heather would both be surprised to know that people send out messages, most of them nonverbal, at the rate of 2,000 to 4,000 on any given day. According to an article in *Town and Country Magazine,* Mary T. Rowe of the Sloan School of Management at the Massachusetts Institute of Technology has coined the term *microinequities* to explain why some people in the workplace feel included and others do not. There is an infinite number of non-verbal communications available to all of us. They might include raising an eyebrow, turning a shoulder, moving too fast or too slow.[5]

We encounter people from time-to-time who are extremely rude, but it's the process, not the content of what's said, that's insulting. If we can't decipher boredom, distain or approval from co-workers' facial expressions or gestures, we're traveling over treacherous ground, indeed. I suspect people who are intentionally rude know this and feel it's safe to hide behind this verbal shield. And they're right. After all, if an employee is rude and you report it to the manager, the first question is always, "What did he say?" We respond, "It wasn't anything he said, but his face didn't look right and he paused at the wrong times and his tone of voice was…ah, forget it."

Reading bodies is just as important, and sometimes more important, than listening to verbal content. I had the privilege of watching the Three Irish Tenors perform and I noticed that one of the tenors, probably a Gatherer, carried himself in a tight, rigid manner while the singer to his right, probably a Hunter, was moving in rhythm with the music, his face glowing and animated. As it turns out, the rhythmic gentleman let nothing inhibit his dynamic personality. He was a physician who had lost both his legs.

As reported by Mark C. Topkin in the *St. Petersburg Times*, Joe Maddon, the manager of the Tampa Bay Rays baseball team, once claimed, "I've never sat down during a game. I think you have to be standing up and moving to think. That's always been the way I do things....I don't like to sit down."

One day, Maddon turned to his spouse when they were talking and asked, "Where's my bat?" She didn't answer and he remembered where he put it. He went to a closet, took a baseball bat out and held it, rubbing his hands up and down its length as he communicated.[6] This seemed to relax him and he was much more at ease.

Some people really do have difficulty thinking without touching something and getting spatial and kinesthetic feedback. Hunters like to stand and move. Sitting at a desk or any confined space can be stressful. Gatherers are comfortable sitting at a desk or a computer. They often find it relaxing.

As mentioned earlier, I was taught as a therapist in training "to listen with the third ear." This means it's important to study the client's physical expression to discover if it's congruent with what is being said and to reveal things that are *not* being said. Words are helpful in conveying information and communicating in general, but they can be used to cover up important thoughts and feelings. The people who rely on e-mail, texting and social sites to communicate are flying in the dark and don't even know it.

Has Superman returned? Can he fly through space and see through walls? Our friend Gary, a Gatherer, is an electrical engineer. He's about to leave a cocktail party on the eighth floor of a downtown condo. He is not familiar with this area of the city, but believes he will have no difficulty finding his car, which he parked on the street a few blocks away. He mentally rehearses his formula for finding his car.

He will take the elevator to the lobby floor and exit to the left to the main doors. When he reaches the city sidewalk, he will turn nintey degrees to the right and proceed to the corner. At the corner he will turn right and proceed south past an alley and an optician's office to the end of that block. Then he will cross the street and again turn ninety degrees to the right.

After he passes a wide alley, he should find the car about halfway between the alley and the end of the street, assuming *he has remembered all the cues in his internal directory.* Of course, now his smartphone has GPS. He'll never get lost and he'll have another machine to show off. IT can help Gatherers with directions and spatial decisions and help Hunters get organized.

Hal, another man at the same cocktail party with Gary, is a landscape designer who will be spending his first night in the city. How does he find his car? Simple. He knows where it is without giving it a second thought. He can picture it in his mind and won't think about it, even when he leaves the party. He'll simply walk straight to it and he might even take a different route back to it, just for the scenic variety and a tiny bit of adventure.

Can Hal see through walls? Almost. Whether he's facing north, south, east or west, it's as though his mind's eye can look through the front or back of his skull, through the walls of the condo building and other buildings, straight to his car. And he really sees it. There, in his mind's eye, is the sleek aluminum body, the rimmed wheels and the leather interior. Hal can see the paint job, even in the dark, and can even smell the leather. Are extreme Hunters and extreme Gatherers different? Yes, absolutely.

As reported in a *St. Petersburg Times* article titled "Lost a car, found a mess," one poor man, undoubtedly a Gatherer, lost his car in a sea of 6,000 parked cars in a shopping mall parking lot just a few days before Christmas. Police saw him wandering around and suspected him of peering into car windows, perhaps looking to steal something. They searched and interrogated him, but found nothing. They couldn't believe anyone could lose his car and spend gobs of time walking around looking for it. Mall security officers refused to comment on the case and the mall management did not return calls, but he was barred for life from shopping at the mall.[7]

This is an example of the other side of the Gather-Hunter coin. The reader may recall our discussion in chapter 9, "Contamination," about Hunter executives who may not exactly remember the fine print on balance sheets and business correspondence. The Gatherer lawyers and accountants couldn't believe it, just as Gatherer school teachers couldn't believe eight-year-old Harvey got the correct answer on a math problem without going through the required sequential steps. Here we have the opposite. Hunter policemen can't believe that anyone could lose a car when it's in plain sight, right there among the other 6,000 cars.

Of course, spatial ability is important for many occupations. Take flying for example. Expert instructors indicate that the more mechanically inclined Gatherers slide right into the airplane groove without a moment's hesitation. When it comes to the instruments, they are reapplying things they already know.[8]

If an artistic type (Hunter) climbs on board, the systems management and machinery factors will take him a while to grasp. Still, when it comes to flying the airplane he'll do as well or better than the guy who knows machinery.

I saw a humorous billboard several years ago. It featured a man with a rigid, ungainly posture, wearing a tight, ill-fitting suit. Even though looking stiff and uncomfortable, he clenched a rose stem in his mouth and gave his best, if unconvincing, shot at doing the samba. Beneath him the headline shouted, "A Passion for the World of Accounting." When Gatherers try to act like Hunters and vice versa, it's a surefire way to get a laugh.

Hazel (a Hunter) and her husband Gary (a Gatherer) enjoyed their ride on a Ferris wheel in Paris and then hailed a taxi to take them back to their small hotel, which was situated near the Notre Dame Cathedral. The taxi driver, who could not understand their pronunciation of the hotel's address, eagerly searched his GPS to find the hotel and get some directions. Hazel looked over, saw the lighted cathedral about a mile away and pointed to it. When they pulled up behind the cathedral, Hazel was able to guide the taxi driver to the hotel. They found their hotel without knowing its address and without the help of an electronic device.

Recently I visited a medical clinic to have a bandage removed. Three stiff-looking people sat immobilized behind their machines (computers). The room was enclosed in glass. I stood there waiting for them to open the sliding glass partition, but they did not look up. Behind them, I spotted someone who looked like a Hunter. He was straddling a chair with one arm over the top and wore the uniform of someone who works with people, an orderly or perhaps even a physician. He actually looked at me…right at me.

I pointed at the bandage, raised my hands palms up and lifted my eyebrows in a questioning manner. Then I pointed down the hall, to the left, in the direction that seemed most likely. He lowered his hands and separated them, like he was signaling a good shot in tennis, smiled and pointed in that direction. About that time, one of the squat robots, whose face was rehearsing a few microinequities for me, including disdain, boredom and inconvenience, struggled to open the smudged, streaked "welcome" window. But I had escaped their clutches and was already on my way.

Comfort in space is a marker for the Hunter. Note that Gatherers can function in space if it includes categorical training, stationary spatial units and sequential rhythm. This is one reason people who have coordination problems enroll in karate classes where they follow rote procedures in a defined sequential manner. Skeet shooting is another example of a controlled spatial situation with external navigational markers. Compare those structured spatial activities with the spontaneous spatial expression found in a boxer or a pro football linebacker.

Gary invites Hal to a baseball game. Because Gary provided the tickets, Hal offers to buy refreshments. Gary has bottled water and a box of candy. Hal orders a large beer, two hot dogs with raw onions and mustard and a bag of peanuts. When they get to their seats, Gary pulls out his Smartphone and reviews the names and statistics of all the players on both teams. He uses the old-fashioned scorecard during the game while getting statistical updates from his phone. He finds two typos and one incorrect batting average on his scorecard. Focusing on his statistical data, he sips some water, but saves his candy for later.

Meanwhile, Hal folds one leg over the empty seat in front of him and begins to wolf down his dogs. A big cigar would hit the spot, he thinks, but he knows Gary would have a fit if he even mentioned it. Gary's hero is a man named Bill James who uses statistical analysis to analyze hitting, fielding and other significant indicators of success in baseball. His work led to the Red Sox winning two World Series after eighty-four years of frustration. Hal is aware of this, but doesn't like to think about it. It just doesn't seem right and after all, it's messing with tradition, similar to the designated hitter.

When the game starts, Gary carefully writes down each and every play on his scorecard and gives a running commentary on each player's stats. Hal ignores Gary's commentary. He knows the approximate stats on the players he follows. This is one situation where Hal excels in recalling detailed information. His sports-wired gender may be overpowering his Hunter personality. Because players are associated with their athletic accomplishments, Hal has no problems with their names or their stats. But mainly, he just loves experiencing the atmosphere at the ballpark.

Gary would give anything to sit in a corporate box high above the field, while Hal prefers sitting with the bleacher bums in the outfield. He savors the camaraderie, the smell of beer and hot dogs and the crack of the bat against the ball. *Poetry in motion*, he thinks. A foul ball heads their way. Gary puts his hands over his head and ducks down behind the seat in front of him. Hal doesn't flinch as he watches the ball sail some thirty yards past them and at least ten yards over their heads.

Hal remarks that Hall is out of his slump and hitting the ball again.

"I think Hall was traded," Gary responds.

Hal gives Gary an incredulous stare.

Gary checks his program. "Oh, you're right. I must have confused him with Huff. The names are so similar."

Hal is dismayed. How could anyone confuse Hall, who is a catcher, with Huff, who plays third base? Didn't Gary know it was the third baseman who was traded and not the catcher? To Hal, these players are not at all similar. Here we can see some clear personality and brain differences in memory. Gary relies on the sounds of names to aid in memorizing lists.

Hal uses contextual clues to help him prioritize and remember important names. Gary does better on spelling tests than Hal and he remembers the names of more players in the game brochure, but Hal is more accurate on the few key players he remembers via use of context. He associates names with size, success, position, temperament, player history and facial appearance.

The game is close until the seventh inning, when the visiting team scores four runs and leads nine to three.

"Let's get out of here," Hal says, starting to stand.

"No, we've got to finish," Gary replies in a stern voice, pulling himself tightly together. "The bullpen has improved several percentage points over the past three weeks and anyway, baseball is nine innings, so we have to see it through to completion. And I have to finish my scorecard and collect the stats for the complete game. It's the only right way to do it."

"Oh, nuts," Hal moans. "That bullpen's nothin' but a bunch of bums, and anyhow, this game is over. No way can these turkeys come back." He gives Gary a crooked grin. "This here ain't the Red Sox, you know." Gary stares at his program, his face hard and impassive.

Hal stands. "Nuts, I'm going for another beer."

13

Interpersonal Politics

Let's have some fun and speculate about art, politics, literature, music and even cooking. We saw that the Gatherers and Hunters of the world mix it up at high corporate levels. Does this happen in the rarified air of television, literature and presidential politics as well? Yes. In fact, as we reach these superior levels of achievement, we may find more examples of extreme personalities.

Take me to your leader.

If you don't agonize about politics, perhaps you should; the mechanization process is infecting our politicians as they react and organize events around them. Many policies are based on polls and partisan compliance rather than values, creativity and leadership. If all we're doing is processing numbers, we are just letting the machines rule us.

Gatherer and Hunter strategies have played important roles in politics. It was rumored former president Bill Clinton's debate team decided to chop diminutive Ross Perot down to child size in order to shrink his bid for the presidency. Prior to a major television debate, Clinton's team rigged the stage with oversized stools, according to a *U.S. News and World Report* article. "When the diminutive Perot sat on his stool, he had to scrunch over to one side just to get one foot to touch the ground. It was designed to make Perot look like a kid, and it worked."[1]

"We also laid out the stage in a grid. We told Bill [Clinton] if you take 10 paces to this point, Perot will be over this shoulder and Bush will be over that shoulder. Bush was caught looking impatiently at his watch, which reinforced the notion that he didn't care about the people who had come to ask him questions."[2] So despite all of the Gatherer focus on content, policies, programs and the rapid-fire exchange of thousands of polished words and phrases, the Hunter nonverbal communications may have won the day.

Remember Richard Nixon's makeup during his television debate with John F. Kennedy? We forget that Nixon won the radio debate with Kennedy, but Nixon's lack of Hunter personality and charisma was no match for Kennedy on television.

George W. Bush was a tad premature when he landed on the deck of the carrier *Abraham Lincoln* to celebrate "victory" in Iraq. But his Hunter appearance was choreographed "even down to the members of the Lincoln crew arrayed in coordinated shirt colors over Mr. Bush's right shoulder and the 'Mission Accomplished' banner placed to perfectly capture the president and the celebratory two words in a single shot. Journalist Elisabeth Bumiller stated in a *New York Times* article that "the speech was specifically timed for what image makers call 'magic hour light,' which cast a golden glow on Mr. Bush."[3]

Historian Roy P. Basler opined that President Abraham Lincoln was a master of public eloquence. The Gettysburg Address was short on factual, Gatherer verbiage and long on creative and mystical Hunter worldview. In fact, Lincoln's speech at Gettysburg didn't even mention the battle itself. The Gatherer professor from Harvard who spoke before Lincoln spent three hours discussing the battle in great detail. Lincoln later said this learned man "can't give a short speech anymore than I can give a long one." At the time of the Lincoln-Douglas debates, Lincoln referred to Douglas as repetitious and his sequential arguments as "interminable memoranda."[4]

Did the lack of IT in the 1860s reduce Lincoln's effectiveness or did it strengthen him? Lincoln was forced to take his time and think things through. The Gettysburg Address was no cut and paste document that could be easily deleted. Rather, it was an inspired message of faith and wisdom.

Recent Republican presidents haven't been long on speeches, either. With the exception of Ronald Reagan, who was a professional actor, Republican presidents haven't showed much verbal ability. President George W. Bush was often criticized for his pronunciation of the word "nuclear," either because he couldn't get it right or wanted to appear as a folksy down-to-earth person. He pronounced nuclear as nu-kye-lar. In an editorial in *The Wall Street Journal*, which criticized computers for going haywire and missing the ten year forecast of economic growth by a cumulative $800 billion, the editors said, "To paraphrase President Bush: That's a pretty severe misunderestimation."[5]

Political columnist Mark Shields reminds us that "Andrew Jackson and Abraham Lincoln...were widely and openly ridiculed as backwoods chowderheads." He went on to admonish Democrats "who have been audibly snickering that Bush is neither the brightest bulb on the tree or the sharpest knife in the drawer, would be well advised immediately wipe the smirks off their faces."[6] Maybe Gatherer-type Democrats and media types equate verbiage with intelligence.

For some Democrats, verbal fluency is next to godliness. Journalist Suzanne Sataline wrote that Caroline Kennedy, in her run for Hillary Clinton's seat as New York State Congresswoman, drew criticism from fellow Democrats for being "inarticulate and vague in her answers." She withdrew from her candidacy "for personal reasons."[7]

Some Gatherers believe in not only looking closely at words, but also examining each *letter*. Jan Morris, author of numerous travel books, who lives in Wales, wrote an article for *The Wall Street Journal* saying, "The letter B, for example, is especially suited to abuse—the words bastard, blithering, brute, bully are just made to be blurted out with blustering bravado. And surely nothing could be more absolutely final than a period—a full stop for the Brits—with no compromising tail or ingratiating squiggle to weaken its decision. Be off with you, you beastly boy, booms the letter B. Right, that's it, clearly declares the period, end of story."[8]

Some Hunters spend a lot of time criticizing the media, which they believe is Gatherer driven. When you arrive at the studio for a television interview, producers or directors tell you exactly where to stand or sit, how and when to turn your head, at which cameras not to look and advise you not to wave your hands under any circumstances. For many panel discussions, you must sit indecently close to another person and keep your voice low. They demand short answers and want to avoid rambling responses.

Maybe this is why some Hunters seem "bug-eyed" and frozen when they appear on television. This is why you see television and radio talk show hosts interviewing other media (Gatherer) colleagues. Instead of going to the source of the story or even to an expert in the particular field of inquiry, they will often interview journalists who have covered the topic. Naturally, Gatherers feel more comfortable with other Gatherers and want to interview people who understand and respect their sequential, logical and chopped up style of discourse. They want people who know the tricks of the trade—let's call them media natives—who are comfortable with robotic exchanges.

Hunters just don't fit into most interview formats. Gatherers approach problems from the bottom up while Hunters work from the top down. In addition, the Hunter wants to respond to the interviewer's questions directly, in their entirety, including the context, whereas Gatherer interviewers have a script and break the subject matter down, piece by piece and sequentially, much to the discomfort of the Hunter. Gatherer interviewers are constantly admonishing Hunter interviewees to wait until a certain topic comes up before discussing it. "We'll get to that later," is a common refrain.

Bill Clinton and John F. Kennedy evidenced strong attributes of both Gatherer and Hunter personalities. Powerful speakers and masters of the linguistic trade, they could also summon humor, emotion and worldview with seemingly little effort. Al Gore and Jimmy Carter seemed more comfortable with Gatherer strategies. They both got caught up in details and couldn't convince the public that their "lust" was real.

A July 2000 article in *Time* magazine reported that presidential hopeful Bob Graham had filled more than 4,000 color-coded notebooks with minutiae such as "watched *Ace Ventura: Pet Detective.*"[9] This Gatherer behavior signaled to Al Gore's advisors that Graham might be poor vice-presidential material. Given the closeness of the race in Florida, keeping the popular former Florida governor off the Democratic ticket may have cost Gore the presidency. This was a mistake of immense proportions.

Thomas Jefferson had similar Gatherer habits and was accused of not comprehending the big picture, according to a *Tampa Tribune* article by Keith Epstein. "Despite recording each tiny expenditure for nearly six decades, Jefferson reconciled his finances no more than two or three times and was a lifelong debtor."[10] Also, in Jefferson and Graham's notebooks there was little sign of emotion. In the same article, political scientist Larry Sabato writes, "Bob Graham's notebooks make me think Bob Graham would be pulling another Jimmy Carter, scheduling the White House tennis courts. Today we psychoanalyze our presidential candidates. We've learned that we need to."[11]

And what would Sigmund Freud say about politician Bob Graham? He might theorize that Graham has obsessive-compulsive traits. His conclusion would differ significantly from the theory of Hunters and Gatherers. Graham might be a moderate Gatherer and left-brainer with some well established, compulsive habits. Some people assume that psychoanalysis fits the Gatherer personality because of its reliance on analysis and talk therapy. But Freud's work also included the study of dreams and the superego, both Hunter domains.

Do politicians with Gatherer personalities tend to represent the political left and politicians with Hunter personalities represent the political right? Do those politicians with *both* Gatherer and Hunter personalities hover around the political center? And do Gatherers become Democrats because they think from the bottom up and focus on individuals who have fallen through society's safety net? Do Hunters become Republicans, because they think from the top down and sometimes miss some of the important "trees" as they survey the "forest" from on high?

We could write this off as just entertaining speculation, but a review of Republican presidents dating back a half century reveals a businessman-pilot, another businessman-pilot, an actor, an All-American football player and a general. Only one Republican president during that period, attorney Richard Nixon, could be classified as a Gatherer, and he was forced to resign from office.

Some Democratic presidents were lawyers and tended to come from the intellectually elite class. Lyndon Johnson didn't come from an Ivy League background, but he was a former teacher. Despite Johnson's effectiveness in the Senate, John Kennedy won the Democratic nomination.

Drew Westen picks up this theme in his book *The Political Brain: The Role of Emotion in Deciding the Fate of the Nation.* Weston believes Democrats tend to be intellectual, thriving on policy debates, arguments, statistics and getting the facts right. He also believes that they are uncomfortable with emotion. One style he discusses is an obsessional personality that involves tone deafness to emotion and "a tendency to focus on details that often leads a person to lose the forest for the trees."[12] Sound familiar?

Some people think Hillary Clinton's presidential campaign wasn't handled well because Mark Penn, her campaign organizer, was, among other things, too data-driven and was not a strategic thinker. In describing Clinton's personality, journalist and author Joe Queenan wrote, "She's just no fun, and politicians who are no fun are hard to write about. A barrel of monkeys is fun. A barrel of dead monkeys is no fun. Hillary is less fun than three barrels of dead monkeys. Maybe 300."[13] Some other Gatherer-type politicians who some feel haven't showed senses of humor include Mitt Romney and Britain's Labor leader, Gordon Brown.

Some Gatherers have poor senses of humor because they're machine driven. Computer "ideas" come stripped of surprises. Have you heard a computer chuckle lately? I can see it now, the computer as a friendly storyteller. "I was tending bar one day, you know, kid, when these two guys came in. I asked the one with the inscrutable expression how the Canadians defended themselves from polar bears. His buddy, a smart-alecky type, says, 'I'll have a scotch with Canadian club.' Not bad, hey kid—not bad?"

The machine has a much greater capacity than humans to memorize and regurgitate stories, but, of course, without warmth or emotion. What the computer can't do is create stories. English political and military leader Oliver Cromwell comes to mind. He not only chiseled saints' names from centuries-old stained-glass windows and engraved tombs, he also "drilled" the term chiseler into our vocabulary. He didn't like people laughing in his church either, although an occasional smile was allowed.

A quick but powerful test for Hunter functioning is to ask a person to read bits and pieces of information about an activity *as the activity is happening* but without seeing the activity, and then announce it as though truly witnessing it. That's what Ronald Reagan did, when, as an aspiring young announcer, he reported Chicago Cubs games from a remote site while reading from a telegraph tape. Listeners thought he was announcing from the press box at the ball park. When the telegraph crashed, Reagan had to call phantom plays and "fill in" descriptions because he had no idea what was happening on the playing field. Sometimes, when the telegraph came back on, the score had changed. Misleading his radio audience became standard fare for the young announcer. As president, Reagan was often criticized for ignoring details and "filling in" from time to time.[14]

Yep, sometimes Hunters exaggerate and "round off," just like they did in school, much to their teachers' chagrin. Reagan never came to terms with evidence that he approved the sale of arms to Iran in an effort to free American hostages in Lebanon. As reported in an article titled "My Heart and My Intentions," he said, "My heart and my best intentions still tell me it's true, but the facts and the evidence tell me it's not." Once he told someone who had reported a harmless but untrue anecdote about him, "You believed it because you wanted to believe it. There's nothing wrong with that. I do it all the time."[15]

Does the prevalence of Gatherer and Hunter attitudes change over time? The Roaring Twenties represented a Hunter era and World War II demanded a Hunter attitude just for survival. The 1950s were a time for Gatherers, a cultural window following the frightening, war-torn 1940s. GI's came back from Europe looking for peace and quiet. They wanted a sitcom life with a small house, a picket fence and children. They'd seen *real suffering* and the results of unrestrained Hunter ambition and creativity.

People who grew up in the 1950s reported it was an unusually good time to live, but if we follow Abraham Maslow's Hierarchy of Needs, people might have regressed some during that time. And some were oblivious to racial justice and other "abstract" causes.

During that two decade period, self-actualization was out and safety and security were in. This was followed by the rebellious Hunter attitudes of the 1960s and '70s. Young men and women demanded individual expression and railed against conformity and "plastic" living. What about today? Will the technology revolution finally put the Gatherer in control for good?

Maybe it already has. During his campaign for president, Senator Barrack Obama received financial support from computer savvy people, many of whom were Digital Natives. People could attend Senator Obama's rallies without charge, but they couldn't get in without giving their e-mail address to the people at the door. Later they were contacted frequently by Obama supporters and many sent in repeated donations in small amounts.

Hunter fiction.

In a summer 2006 editorial in *Phi Beta Kappa*, editor John Churchill criticized Tom Wolfe, author of *Bonfire of the Vanities* and other bestsellers, who spoke at the annual Jefferson Day Lecture in Washington, D.C., for "odd," "anecdotal," "confused" and "unconnected" statements.[16] Churchill does not criticize the *content* of Tom Wolfe's remarks, but criticizes Wolfe on his style and process of presentation. Wolfe's style, it so happens, is a textbook example of creative, Hunter-type thought processes.

Churchill further stated that Wolfe needed a more *recursive* approach. This means "the determination of a succession of elements by operation of one or more preceding elements according to a rule or formula involving the use of a finite number of steps."[17] This is heavy for anyone with even a few Hunter genes. It's a good example of Gatherer sequential processing, where each new comment becomes the subject matter of its successor. Could we boil this critique down to a Gatherer (Churchill) chastising a Hunter (Wolfe) for being what he is—a Hunter? Maybe, maybe not. Hopefully, future research will tell us. In the meantime, Mr. Wolfe is still on the Hunt, writing great American novels.

Hunters and Gatherers definitely write differently. Author Nancy Kress, in an article in *Writer's Digest*, described some of the different options for *re*writing fiction. I'd put her first group in the

Gatherer arena. These folks write constantly as they go along and then stop and rewrite on the go. The second group of writers produces a *chunk* of fiction and then revises it before going on. Remember how Gail, the Gatherer, helped Hal, the Hunter, in chapter 7 by organizing his chores into chunks? The third group, Kress states, "writes the entire first draft like a person fleeing a bear: Go fast and don't look back. The first draft may end up in a holy mess, but at least it's something to work with in the rewrite."[18] No question in which personality group this last type of writer fits.

The eye of the beholder.

Wouldn't we expect art to fall, hook, line and sinker, into the Hunter arena? Not necessarily. Gatherers, if they like art at all, prefer realistic art that is almost photographic, while Hunters, especially extreme ones, prefer impressionistic art and have a soft spot for Salvador Dali's creative subconscious and Dale Chihuly's glasswork. When Gatherers view art they prefer to stand close where they can see the details, including brush strokes, while Hunters observe from a distance to get a better overall impression. Even museum docents differ. Gatherer types will emphasize symbolism, the history of the painting and the biography of the artist, while Hunter docents will focus on light, color, mood and how the painting affects the viewer emotionally.

Artists have to avoid over-control of their paintings. Sometimes they need to get out of the way of their own paintings and let it happen! This is a chronic struggle for artists, because the Gatherer side keeps interfering and trying to add control, sequencing and other mechanistic techniques learned from the machine culture. Sometimes artists learn various drip techniques that allow the paint to drip freely down the canvas, creating spontaneous images. These techniques give the artists permission to let go.

So far the computer hasn't morphed into an artist, so we've got no competition, right? Hold it a second. What about chimpanzees? Psychologist Michael Gazzaniga says in his book *Human: The Science Behind What Makes Us Unique*, they'd rather paint than eat and at least one of their paintings was sold for $20,000. They reject dull crayons and throw them at researchers with the justifiable indignation of an artist.[19] Fortunately for us, and our computer friends, they can't produce a recognizable image.

In the highly regarded *Barnes Institute* in Philadelphia, paintings are not hung sequentially because Barnes believed that "artists don't think that way." He's right. I talked to an artist who contributed works to an upscale craft show and he reported that most people who attended were quite deliberate in their assessments of the work prior to making a purchase. But the artists were astounded to note that an accountant had asked her friends, who had already seen the show, to assess the quality of pieces on display. When the Gatherer arrived, she had a list of items with her. She just purchased each item, without really looking at it and checked it off her list.

These differences between Gatherers and Hunters are seen in music and entertainment as well. Hunters prefer sitting at the front of the concert where they can see the musicians' movements and facial expressions whereas Gatherers are comfortable sitting farther from the stage, where they can get a better acoustic experience. Speaking of music, a piano instructor once told me, "At the beginning, students play from the rules (Gatherer); later they play from the heart (Hunter)."

Three Hunters—a painter, a landscape artist and the owner of an art gallery, attended a mystery club party. A crime scene was described in detail and there was lots of dialogue. It reminded these three Hunters of mystery novels, which they hold in great disdain. When the goal of the game was announced, they couldn't believe it. The goal wasn't to identity the murderer, who was rather obvious. Instead, the winner was the one who identified the greatest number of clues.

The Gatherers were into the game big time, but the Hunters were bored and tuned out. They identified none of the clues and instead discussed new scenarios with fantastic outcomes. They got so carried away with their revisions of characters and events that they incorrectly identified the killer. The facts, details and sequential dialogue reminded them of learning phonics in grade school, not good memories.

How about cooking? Gatherers can turn out a fine meal, but it takes them a long time and they don't dare deviate from the computerized recipe. When they serve the food, each dish individually is quite good, although perhaps lacking appeal. The problem comes in the balance and integration of the various dishes and seasonings. They're thinking of one dish at a time, rather than, for example, serving a gelatin salad to cut a rich and creamy pasta dish.

As good Gatherers, they're working from the bottom up rather than the top down and sometimes miss the big picture. The Hunter, on the other hand, rejects recipes *and* computers. He or she cooks creatively by taste and visual presentation, integrating dishes. The Hunter may overlook some vital detail, however, leading to disastrous results.

14

Machines vs. Humanity

If the growing epidemic of machines infects us all, I believe we'll lose our humanity. And our humanity is heavily invested in art, religion, philosophy and the creative aspects of our lives. Once again, let's look for significant differences between Hunters and Gatherers.

Angels and astronauts.

Science and religion are not usually happy companions. Science protests that religion is not provable (unscientific) and religion complains that science is limited (ungodly) because of logic and reductionistic thinking. Sounds like the Gatherers and the Hunters are at it again!

Remember, left-brain and right-brain research provides the neurological underpinnings of the Hunter/Gatherer theory. Psychologist Robert Ornstein helped us understand how the two sides of the brain perceive things differently and I believe his explanation is useful in understanding some of the differences between science and religion. In his book *The Right Mind: Making Sense of the Hemispheres*, Ornstein states, "The two sides (of the brain) handle the world differently, one focusing on the small elements of a world view and linking them tightly together so they can be acted upon, produced, reproduced, like a formula. The other links together the large strokes of a life's portrait, where we are, where the parts fit, the context of our life."[1] Once again, we see Gatherer formulas versus large Hunter strokes.

The Futurists met in Philadelphia in 2002 for an annual meeting. Futurists try to look at four things: What is probable, what is possible, what is preferable and what is preventable. Futurist Joseph Coates mapped out the future for the next 1,000 years, saying, "Futurists imagine a world with no crime and where people will use genetic engineering to raise the average IQ to more than 180. They will give up religion. Earthlings will learn to avoid asteroids and they will eschew meat for protein powders."[2]

Theologians and philosophers fear that Gatherer scientists are destroying the very essence of the human worldview. In his book *Heraclitean Fire: Sketches from a Life Before Nature,* biochemist Erwin Chargaff says, "The wonderful, inconceivably intricate tapestry is being taken apart strand by strand; each thread is being pulled out, torn up, and analyzed; and at the end even the memory of the design is lost and can no longer be recalled."[3] And theologian Martin Luther stated, "For reason is the greatest enemy that faith has: it never comes to the aid of spiritual things, but—more frequently than not—struggles against the Divine Word, treating with contempt all that emanates from God."[4]

Even within the religious community we find Gatherers and Hunters offering quite different perceptions of the spiritual world. The bumper sticker, "My Karma Ran over Your Dogma," says it all. Apparently some Hunters don't want to be tied down by Gatherer religious doctrine.

If mechanistic thinking begins to dominate our culture, will that be the end of religion? Many scientists are poised to put "mythology behind them" and move on to a brave new world. And they'll have support from a number of entertainment and media types. In a *St. Petersburg Times* interview, performer and philosopher Bill Maher (*Real Time with Bill Maher*) stated, "The folks in this country who believe in science and rationality as a means of governing over faith are left without a choice. Somehow, John Kerry, Al Gore, they all felt the need to buddy up to the red-state, NASCAR, gun-loving, beer-drinking culture." He went on to say,

"I wish we had a candidate who said, 'You know what? My religion or whatever it is, is none of your business, and it doesn't really affect how I do the job of president and I'm going to promote science and rationality—how about that?'" He comments that religion is a neurological disorder. "You know how religious people say about the gays: 'Hate the sin but love the sinner?' It's the same thing here: I hate the neurological disorder, but I love the people who suffer from it, and I want to help them."[5]

As documented in the article "A photo album of the soul" by John Barrow, Dr. Andrew Newberg studied Tibetan Buddhist monks, cloistered nuns and Pentecostals who speak in tongues. Using radioactive isotopes, he found that the nuns and monks showed excitation in the frontal lobes and thalamus and they had *shut down* part of the parietal lobes.[6]

Now that MRI machines are helping neuro-scientists release journal articles at a faster pace than any time in history, some scientists theorize that religion and music are just electrical firing. Maybe they think brain synapses are having a meeting at the *old revival tent* in our brainstem! In 2006 Jorge Mall and Gordon Grafton, neuroscientists at the National Institute of Health, showed that volunteers undergoing brain scans who placed the interests of others before their own interests activated a primitive part of the brain usually associated with food and sex.[7]

Since food and sex are *mostly* pleasurable, I suppose when we help others, we're just satisfying our own basic needs. Does that mean that people who are obsessed with sex are saintly? Or do I have this thing turned around? If it does, I guess the 1960s' cry of "make love, not war" finally makes sense. The researchers go on to say that altruism is not a superior moral faculty that suppresses basic selfish urges, but rather is basic to the brain, hardwired and pleasurable.

Aren't some brain areas associated with everything we do? How does pinpointing an area in the brain associated with altruism add to our knowledge? How do we go from that finding to declaring that altruism is not a moral force? This research puts us no further ahead than we've been for centuries. In addition, we know that the brain has plasticity and can change in response to the environment and our experiences. Maybe our altruistic choices helped fortify the sweet spot for altruism and not the other way around.

Some scientists are eager to jump on these kinds of findings and infer that there is nothing special about human morality and human potential. It's just another hardwired product of evolution.

Biologist and theorist Edward O. Wilson, in his book *Consilience: The Unity of Knowledge*, states that the brain is a machine and the mind is the brain at work.[8] Cultural and economic critic Wendell Berry calls this the Tarzan theory of the mind and questions whether a human raised entirely by apes would still have a mind. Berry comments on the scientific, technological and industrial components of medical science in his book *Life Is a Miracle: An Essay Against Modern Superstition*. He says, "Our daily lives are a mockery of our scientific pretensions. We are learning to know precisely the location of our genes, but significant numbers of us don't know the whereabouts of our children."[9]

Some Hunter types seem to agree with Berry. They argue against philosophical materialism's stance that values are only what can be seen and measured. "The tyranny of mechanistic thinking has also affected contemporary social policy and moral discourse," says essayist Jeffrey Hammond in his article "Lost Souls," published in the *Notre Dame Magazine*. "At root, the denial of the soul reflects a misguided rage for order. It is an attempt to assert control over what cannot be controlled. But if we trust only what we can observe and predict, we will have an excessive fear of the random, the unusual and the abnormal." He goes on to say, "Anthropologists and psychologists tell us that we are not born knowing how to be human. This has to be taught and learned. The soul may well serve as ethical and metaphysical shorthand for our capacity to absorb this complicated bundle of lessons in the course of becoming fully human."[10] Note Hammond's reference to mechanistic control and order. Sounds like the Gatherer in us.

"If we accept that knowledge is a finite island in a sea of inexhaustible mystery, then two corollaries follow: 1) The growth of the island does not diminish the sea's infinitude, and 2) the growth of the island increases the length of the shore along which we encounter mystery," says writer and professor Chet Raymo in an article in *Notre Dame Magazine*.[11]

Concerned about too many details, too much prioritizing and excessive multitasking? Maybe an antidote can be found in the New Testament. Jesus entered a village where a woman named Martha welcomed him. Her sister Mary sat beside Jesus at his feet, listening to Him speak. Martha was burdened with all the serving and asked Jesus, "Do you not care that my sister has left me by myself to do the serving? Tell her to help me."

Luke 10:38–42 describes how Jesus replied to her. "Martha, Martha, you're anxious and worried about many things, when there is need of only one thing. Mary has chosen the better part, and it will not be taken from her."

Getting back to the neurological underpinnings of the Hunter-Gatherer theory, it's interesting to note that knowledge of hemisphere interaction may predate modern discoveries based on split-brain research. My review of psychological literature shows theoretical systems virtually identical to brain function as we know it today.

Psychiatrist Eric Bern's theory of transactional analysis proposes three ego states: Adult, Parent and Child. The Adult is rational, provides good data, *gathers* information and is counter impulsive (read left brain and Gatherer). The Parent deals more with the ideal than the factual and draws on traditional, worldview perceptions (read right brain and Hunter). The Child is closer to instincts and basic physiological processes (read lower-brain centers).[12] Does current split-brain research offer scientific validation for Bern's theory?

Even before Bern, another major theory is consistent with today's neuropsychological findings and our Hunter-Gatherer ideas. Sigmund Freud proposed that the personality is made up of three major systems: the Id, the Ego and the Superego. The Id is the reservoir of psychic energy and furnishes all of the power for the operation of the other two systems (read lower brain centers). The Ego has control over the cognitive and intellectual functions and obeys the reality principle (read left brain and Gatherer). The Superego represents the ideal rather than the real and focuses on the traditional values of society (read right brain and Hunter). According to Calvin Hall and Gardner Lindzey in their book *Theories of Personality*, "Behavior is nearly always the product of an interaction among these three systems; rarely does one system operate to the exclusion of the other three."[13]

In the religious concept of the Trinity, I believe God the Father corresponds to the Adult, Ego or left-brain Gatherer who lays out the commandments and dogma of the church. The Holy Spirit is the Parent, Superego right-brain Gatherer, speaks in tongues of fire and inspires the followers of the church. The Son of God corresponds to the lower-brain centers, brings rebirth, renewal and energy and is God-made-flesh.

The concept of a Triune God is difficult to grasp, although philosopher and theologian St. Augustine and a few others made a valiant effort to explain how three persons could function within one nature. Could Wendell Berry, Erwin Chargaff and Martin Luther be wrong? Has science's insistence on analysis and reductionism finally helped to confirm a fundamental religious belief? Rather than an enemy of faith, could science now help us to better understand the broad fabric of the worldview tapestry?

And could ancient monks, without the benefit of MRIs and other imaging technology, comprehend and develop a spiritual *and* scientific truth? What does this mean? Did man create religion based on a subconscious awareness of these brain functions or did God reveal an immense and infinite mystery that until now was too complex to fully comprehend?

In writing about cooperation between religion and science in *Life is a Miracle*, Wendell Berry could be describing the two sides of the brain and the competition between the Hunter and the Gatherer personalities. "Both *imagination* and a competent sense of *reality* are necessary to our life, and they necessarily discipline one another. They should cease to be 'two cultures' and become fully communicating, if not always fully cooperating, parts of one culture."[14] (my italics)

When Eve took a bite out of that cold, metallic apple, perhaps it gave her Gatherer ideas about knowledge. Until that time, Adam and Eve had been simple, intuitive and childlike, which sounds like Hunter characteristics to me.

Our uniqueness shows itself in our behavior, although some individuals are more extreme in expressing this. The incredible documentary *Man on Wire* recorded high wire artist Philippe Petit's successful attempt to walk on a wire between the roofs of the two former towers of the International Trade Center. He not only walked, but also occasionally took a nap while lying on the wire some thousands of feet above the ground. Later, when Petit was asked why he'd be motivated to do such a thing, he said simply, "I needed to touch the clouds. In life you must refuse the rules and refuse to repeat yourself."[15] I don't think the repetitious and controlling nature of machines would appeal to this man.

Perhaps like the Hunters, Gatherers are also looking for manna from heaven. In the article "The Singular Question Of Human vs. Machine Has a Spiritual Side," columnist Lee Gomes interviewed a group of people concerned with "singularity," a word that refers to the coming of a machine synthesis into a new, super intelligent life form. Gomes suggests that the discussion of singularity might involve a sublimated spiritual yearning for some form of eternal life. Gomes quotes a poem read at the singularity conference that describes an Aquarian age scene in which humans and other mammals frolic in a cybernetic meadow, watched over by machines of loving grace. He states, "Those computer protectors sound a lot like the guardian angels my grade-school nuns told us about."[16]

Illustration courtesy of Dave Coverly

"That's right, Danny, the stork used to bring babies, but now mommies can just download them at stork.com."

15

Future Mechanization Scenarios

If you think machines infect our minds and spirits today, the future holds the promise of even greater pandemics. On a frosty, unseasonably cold September morning in 2071, Strike Force One, an elite corps of orbiting commandoes, landed at Bantry Bay, Celtia Union. Celtian farmers had been draining oil from U.S.S. pipelines in the Atlantic Ocean, near the former country of Ireland. The U.S.S. Scientific Council ordered a controlled military strike to teach the Celtians a lesson, but the strike force of 874 soldiers ran into a snag soon after landing. A computer snafu ordered the diminutive commandos to exit their space ships and then return to them within minutes. This was followed by a steady stream of orders that had them exiting and reentering their ships continuously for the next two days.

Finally, Celtian soldiers took pity on the frail and exhausted commandoes and severed their communication lines. Believing they had been delivered from these invaders, the Celtians were in a festive mood. They paraded their visitors through grassy woodlands and pastures to lovely shaded villages where they offered them spicy foods and foaming beverages in brown bottles. Voluptuous women with pendulous breasts embraced them and danced through the streets.

Stunned commandoes watched the brawny, meat eating Celtians brutally, and happily, knock one another down while attempting to kick a ball through posts at each end of a field. These tribal customs amazed them. Offers of marriage frightened them even more. They didn't know what the word meant. Was this the same as marrying two procedures in a scientific experiment? When the Celtians finally put them back on their ships and sent them home, the commandoes were damaged beyond repair and had to be deleted (killed). Some had stayed in the Celtia Union, but the returning soldiers showed symptoms of post-traumatic stress. Weeping and other signs of emotion terrified them.

I was one of those soldiers. My name is Integer 372-Zeta Max 32420. The year is 2084. I write to you from Darwin, District of Colorado, Capital of the U.S.S. (United Scientific States—formerly the United States of America). What you are about to read is factual. This is not science fiction. It was outlawed in 2068. Zeta is the name of my robotic peer. We are friends. This will be a brief history, as I am writing from the Energy Source Laboratory, deep in the catacombs of the R1 National Science Center.

As a nation, we are ahistorical. Writing about the past is prohibited. Our belief is that every day presents a new opportunity for exploration and discovery, unhindered by historical ideology or mythology.

Changing the name of the country and moving the capitol to the world's largest research laboratory, here in Colorado, was first proposed in 2060 when Canada, Western Europe and the Netherlands petitioned to join the United States. The transition from a democracy to a scientific council made up of America's brightest scientists wasn't completed until 2066, but historians trace its roots all the way back to 2008.

The unprecedented financial crash of 2008 led to a decline in public faith in private and governmental institutions. Research shows that the government and Wall Street knew about and contributed to the prevalence of bad debts that would bring down America's greatest investment banks. The reigning ethos became "every person for himself or herself." As a result, our leaders believed greater scientific authority was needed. Scientists were also outraged at obscene bonuses paid to high-level CEOs and the income earned by low-IQ uneducated people, who ran small businesses.

This same period also brought a dramatic change in religious values. For the first time, a major democratic government authorized the killing of human life in order to potentially improve the lives of living persons. Not surprisingly, later in 2008, religious leaders objected. The Vatican issued *Dignitas Personae* (the Dignity of the Person), denouncing attempts at futuristic possibilities such as cloning people using gene therapy to enhance the human race.

While much of the focus at that time was on stem cell research, the major breakthrough came from the government's willingness to allow the scientific community to regulate itself without outside interference. Scientists formed a council of the brightest and most accomplished researchers. This assembly became the predecessor of the U.S.S. governmental structure.

German scientists deciphered the genome of Neanderthals. Since Neanderthal and human genes are mostly the same, scientists at Harvard created a living Neanderthal by splicing Neanderthal genes into a human genome and using chimpanzees as surrogate mothers. Religious leaders complained and wanted to know if the clones should be offered professorships "in Harvard or in a zoo"[1] as said in a *New York Times* article by Charles R. Morris.

Scientists gradually imposed their own vision of world progress, which led to rule by scientists and intellectuals. In 2066, scientists gained control of the federal government. The new High Commission on Science rejected traditional value systems and also rejected logic as regressive. Scientific leaders finally determined that science itself was a value system and defined its goal as the need to expand the limits of the human condition.

In the early days, the Scientific Council selected the brightest children and assigned them to schools of science to learn scientific methodology, research design and statistical analysis. Defective specimens were deleted or sent to research laboratories. Defectives, if they lived, were prohibited from procreation.

Surrogate motherhood became a class and gender minefield. The going rate for exceptionally endowed surrogate mothers was about $25,000 in 2010 and, adjusted for inflation, reached $50,000 in 2015. In 2072, families were asked to contribute their progeny for research purposes. They were given much praise for this Darwinian act of species enhancement and honored as heroes. A flag was placed in the front window of each of their homes with the number of stars indicating the number of units contributed. The research programs were kept secret, not because they might upset parents or others in the community, but because there might be some objection to the research design or statistical analysis.

It soon became apparent that perfect human specimens were necessary to fulfill the state's goals. Scientists speculated that selfish impulses led some parents to keep the best specimens. Other parents surrendered to the nagging problem of emotional attachment. As a result, children were chosen randomly on Life Day, once a year. However, longitudinal research studies raised questions about this selection process.

Why not begin experimentation in the womb and during the first hours after birth? Utilizing knowledge of the plasticity of the brain, those brain cells needed for scientific inquiry would be bombarded with stimulation based on electronic game research. Why have to recondition the brain and alter its chemistry when one could "grow" the proper brain to begin with?

Some of the few remaining historians recalled deeds of brainwashing during the Korean War and the media reviewed famous crime cases in which brainwashing was supposed to have taken place, such as the kidnapping of newspaper heiress Patty Hearst. But no one knew exactly what procedures were most effective. Now the state could truly enhance its citizens' cognitive abilities for their own good and the good of the state. With alteration of brain cells taking place in adults over a short period of time, scientists were excited to begin working on the soft tissue of fetuses, infants and young children.

At that point the scientists had a potentially achievable scientific goal. But how would they produce a Mozart or an Einstein or a Bill Gates? Research on these rare specimens theorized that: (1) If a child is pre-wired genetically, (2) has a marvelous educational exposure, (3) has the motivation, stamina and good health to practice for thousands of hours and (4) doesn't mind being different from others—presto, a genius is born! And now, for the first time, they wouldn't differ from others. Most would be geniuses.

A secret weapon practically guaranteed the proper result: electronic game technology. After fertilizing wombs with sperm from leading scientists, they bombarded the areas of the infant's vulnerable brain while still in the womb and during the first months and years of life with stimulation of brain cells needed for scientific excellence, while simultaneously shrinking areas that housed philosophical and emotional-based ideation.

This cerebral manipulation helped the government battle against underground remnants of a small but belligerent core of religious zealots. Religion, philosophy and American historical figures were problematic, because the media used them to oppose scientific progress at every turn. To further discourage mythology, civic festivals to honor famous historic icons were eliminated and replaced with scientifically labeled holidays such as Darwin Day and Cyber Monday. The only holidays for schoolchildren were scientific field trips within the U.S.S. Trips outside the country might expose the children to mythological thinking.

Santa Claus was condemned in 2074. Children needed to be protected from foolish superstitions and confront facts. Art and fiction were also on the forbidden list. Science fiction was acceptable until 2068, but research conclusively demonstrated that it had a debilitating effect on readers and encouraged subversive criticism of legitimate science…and the state.

The government curtailed Internet access in 2073. Research showed that Internet postings from the Celtia Union and Third World countries reinforced emotional and mythological thinking. Every person in the U.S.S. was assigned a personal robot or avatar. These robots were numbered and exchanged frequently to avoid a state of dependency, which could lead to unhealthy (emotional) attachments.

Initially, the legal profession was instrumental in enforcing scientific statutes and regulations, but when full scientific control was finally established in 2075, lawyers were no longer necessary. It's hard to believe now, but taking a lethal dose of pills in front of a live Internet audience with people urging the person on, caused a minor stir back in 2009, but it seemed to lose its drawing power as it became more common and the population lost its capacity for emotional response.

The government initially supported suicide because, de facto, it helped eliminate inferior specimens. The body of law concerning duty to rescue the injured was also eliminated. There was no longer a requirement for bystanders to become Good Samaritans. When the suicide rate increased dramatically in 2071 after the unsuccessful raid on the Celtia Union, the Scientific Council made suicide illegal. Persons caught in the act of suicide were deleted by the R.E. (Response Extinguishers), a special police unit. Later, the council decided to save suicide-prone individuals in order to learn more about their character deficits.

In the 2070s, research showed that scientists were coming up with fewer creative ideas. Some council members postulated that the programmed shrinking of areas of the brain associated with emotion had curtailed creative abilities. Emotional intelligence dwindled, and people became a caricature of intellectualism and high IQs. They were adept at logic and scientific thinking, but inept in the interpersonal world. Predictable, critical, fastidious, inhibited and uneasy with sexuality and sensual experience, they were detached and emotionally bland.

Many females became anorexic and nonsexual. There was teasing about a mythological princess back in the dark ages (known only as Marilyn M.) who had walked in a way that stirred men's passions. Fluent in expressing their thoughts, females became introspective, anxious and expressed anger indirectly. Men and women held each other in contempt. In 1965, according to some research studies, 85 percent of our citizens developed a meaningful philosophy of life, but this had dropped to 5 percent by 2084. Financially well-off persons grew from 28 percent to 94 percent over that same period of time.

Today, U.S.S. citizens have become ahistorical. They take no lessons from history and do not recognize responsibility to the past. Slower to emerge into adulthood, they show more dependence on parents and quickly regress when challenged. Research using the Narcissistic Personality Inventory indicates an 84 percent rate of narcissism in 2084.

Laws passed in 2078 replaced pets with robotic toys. This was for hygienic reasons. These robotic cats and dogs responded to verbal commands such as "sit" and "lie down." They blinked their eyes wide, wagged their tails and sniffed before chewing on a bone. Just two years later, in 2080, these toys were banned because some individuals were becoming emotionally attached to them.

All behavior was classified as a function of stimulus and response. Creativity was strictly the product of appropriate reinforcers provided by society. Are our citizens happy? As emotion is not rewarded, the average citizen cannot understand this question. Most feel they are in a state of continuous partial attention and somewhat numb to the world around them. The emphasis is on being efficient and productive, so as not to face deletion.

By 2084, the other major world power is the Celtia Union, which is made up of Africa plus the former countries of Ireland, India, China, Argentina and Brazil. It is a democracy, similar to the former European Union. Family cohesion is a prominent, civilizing force and various religions still prosper. Christian, Jewish, Muslim, Hindu and Buddhist holy days are celebrated. Most Christian families still enjoy and cherish the excitement, gifts and magic of Christmas. They believe it is dishonest and abusive to reduce Christmastime to the scientific analysis of the gravitational problems of a sled and eight tiny reindeer.

The U.S.S. has declined in population to ninety-four million. With the use of cloning, reproduction is authorized primarily for research purposes. The Celtia Union has grown to six-billion and the average family has about three children (no one knows for sure, as statistics aren't kept). Today, the average citizen in the U.S.S. lives to the age of 136. This average would be significantly higher if it weren't for the increased suicide rate, which spiked after the infamous Celtia Union raid in '71. The average life span in the Celtia Union is about ninety-seven—but no one knows for sure.

In the Celtia Union, private property and privacy are protected, as are nature and wildlife. The economy shows moderate growth, with spikes and depressions from time to time, and is based on small, family-owned businesses and agriculture. While businesses in the U.S.S. are based on a scientific and behaviorist model, small businesses in the Celtia Union succeed because they are not bound by the measurable. They exploit family ties, high levels of motivation, creative thinking, group loyalties and trust.

It is difficult for the few Hunter types (such as myself) who somehow survived cortical "repair" and early educational brain restructuring, to work and survive in the U.S.S. Because of my spontaneous creative thinking, which is not based on scientific principles, my family and I, including my robotic peer, Zeta, are branded as damaged or treasonous. I have no way to display my creative personality and cannot afford most of the benefits of technology.

Others like me, whose families migrated to the former United States in the eighteenth century, have now begun to reverse their migration from the U.S.S. to the Celtia Union, where there is appreciation of artists and creative thinkers. This is a risky business, however. If my escape plans are discovered, I will be deleted by the O.U. (Outlier Unit), because it is feared I might harbor scientific secrets.

Needless to say, the U.S.S. Scientific Council has seriously clung to the goal of obliterating the Celtia Union, especially after the Celtia Union Raid in '71, but agricultural supplies are desperately needed and there isn't a sufficient number of strong, well-coordinated U.S.S. workers to run the Celtia Union farms.

They have also hesitated because intelligence services report unusual and creative activities from time to time. The U.S.S. has developed computers that can master any human in the game of chess, but the intelligence corps discovered that a few people in the Celtia Union could defeat the U.S.S.'s latest model computer. If they could do that, what else might they be capable of doing?

Just the size of the Celtia Union was another problem, and its citizens had put their country first during several military clashes with other countries. This blind patriotism made them dangerous and impossible to understand. While the U.S.S. has advanced scientific weaponry, it fears some creative military response on the part of the Celtia Union. Who knows what these odd, whimsical people might conjure up to damage the U.S.S.?

The intelligence corps also reported back in 2069 that the Celtia Union might have had contact with visitors from another planet. The U.S.S. has invested 40 trillion dollars in a research project to make contact with space aliens but so far has not succeeded. This worries scientists. Why did people evolve only on this tiny planet? Some wondered about a *causative agent*, but this was hushed up and hidden from the dreaded media.

Later investigations revealed these visitation rumors were based on a superstitious belief that a visitor from outside the planet, disguised as Jewish, had visited Earth some twenty-five-hundred years earlier. Still, these emotional and logic-based people were erratic and unpredictable. A cold war between the U.S.S. (Gatherers) and the Celtia Union (Hunters) has now existed for more than forty years.

We must go now. Zeta (my robotic peer) and I are on the run. If we make it to the Celtia Union we will teach this history to others.

Goodbye. Integer 372-Zeta Max 32420.

16

From Impersonal Contact to Reconnection

Will we evolve into a U.S.S. or Celtia Union? No. Not if we protect ourselves from infection and prepare ourselves by careful measures, studying the formation of pandemics. The best-case scenario will give us the benefits of IT without losing our humanity. Our control over the use of machines will give us more time to enhance our relationships and self-understanding.

Bringing families closer together.
Currently available controls allow parents to prevent their kids at certain ages from logging on to the Internet. When Garrison, a thirteen-year-old, tries to log on, the message is clear: "You can't log on because of parental controls." Parents must control, not enable, their children. Censorship? No, just parents fulfilling their responsibilities to their children. This is no different than sending children to bed at a certain time each night or parceling out allowances and household chores. Parents need to be familiar with and understand the technology in their homes and engage in game play with their kids. You can't fix something you know little about. Family game activities, electronic and face-to-face, should be emphasized. The family needs to do media projects together.

Families should place computers and game stations in hallways and kitchens, etc., where they can monitor what their kids are doing, at least most of the time. Teenagers with cell phones should be required to subscribe on an individual basis. Even if parents are paying for phone usage, their youngsters will have a better appreciation of the cost of phone calls *and texting*. This does not happen when family phone plans are used. The number of calls should also be restricted and adjusted according to circumstances. If properly used, cell phones can enhance family cohesiveness. They let parents know the whereabouts of their children and provide a safety net.

Parents need to use electronic games as a reward rather than as a *privilege*. Usage needs to depend on appropriate behavior such as study time or cooperation with siblings. They shouldn't be used after displays of negative attitudes or behavior. Parents must follow the Entertainment Software Rating Board (ESRB) guidelines for age-appropriateness of computer and game content. New guidelines for *process* as well as *content* need to address the vulnerability of the young brain. Exposure time to basic games that do not enrich the brain should be carefully limited even if they are not violent in content. A balanced approach to scientific (Gatherer) and creative (Hunter) thinking must be allowed to evolve.

The solution to the potential negative impact of electronic games may eventually come from the industry itself, as it responds to consumer safety awareness. In 2008, stores began to stock less violent "family fare." According to an article in *The Wall Street Journal*, "Game makers such as Sony, Electronic Arts Inc. and Microsoft Corp. have poured millions of dollars into family-friendly fare."[1]

Improving game *content* is a big step in the right direction, but it's a completely different challenge to assess which games are not activating important brain centers. Games that don't activate the frontal lobes and working memory could starve important areas of the young brain. Not every game or activity needs to stimulate the brain in an important way (we all deserve a break from time to time), but overuse of gaming by young children could be a significant problem. The question is not just which game should be played, but for how long.

Parents need to lead by example. As role models, they need to use electronics in a mature and appropriate manner. They must limit the amount of time that they are using their machines for business or pleasure and refrain from using them while driving or in other situations that require concentration on two or more tasks at one time. Ito's research group found that about a third of parents reported playing electronic games with their kids and believed this helped bring their families closer together.[2]

Warm apple pie in the old school house.

Art, music and outdoor play need to balance rigorous teaching of the three R's. Classes for the responsible use of IT would update students on the benefits and drawbacks to learning and self-esteem. Cell phones should go into the "hot box" at school. This form of censorship should be consistent with current restrictions on potential weapons and inappropriate dress and grooming that distract students from learning.

Parents and teachers need to work together to gradually expose younger children to the broader reaches of IT as their brains mature with age and appropriate exposure. The physical damage to young children who are exposed to heavy weightlifting is an appropriate analogy. Another is obesity. Adults who occasionally consume fast foods are much less vulnerable to obesity than are children who grow up consuming them on a regular basis: fast food natives versus fast food immigrants? Future classroom discussions of appropriate use of social networking sites should include research on privacy and cyber bullying.

In the future, computer and electronic game companies need to work with educators and neuropsychologists to develop instructional games based on the "Best Practices in Teaching" compiled by Professor Tom Drummond.[3] Congress needs to authorize the expenditure of trillions of dollars to create a national pre-kindergarten through high school electronic curriculum to *supplement* traditional classroom teaching. This smart software would assist unique students with distinct learning styles and team-based learning will utilize digitally equipped classrooms.

We must develop other innovative approaches that will assist students. Smaller schools offer a more intimate and personalized approach to learning and maturation. We must stop the huge factory-machine atmosphere of schools with more than 500 students. I believe credits for completing specific course work should replace the present annual progression from "grade" to "grade." The expansion of charter schools, vouchers and home-schooling would allow for more innovative learning and assist low-income students. Busing should be reduced and remaining school buses need to be converted to rolling classrooms utilizing online courses and video learning.

To assist children with learning deficits and other developmental problems, we have to utilize highly sophisticated and individualized profiles of strengths and weaknesses. Machines can assist with *some* of the teaching through the child's strong and open avenues of learning while well-trained and sensitive remedial teachers give the child one-on-one instruction in weak processing areas.

Creative expression.

Once individuals recognize the dangers of IT—as well as its blessings—towns and communities need to set aside green, open spaces to encourage free play and face-to-face game playing. Movements such as Creative Cities International (CCI) believe that culture and community are at the heart of urban planning and declare that "to be successful, cities must not neglect their creative and cultural potential."[4]

In the future, integration of intellect and emotion must become a key societal ambition. Not only should game developers drop the basic coded games and restructure them as artistic mediums, but parents need to inoculate their kids from machine-born pestilence through exposure to Hunter-type activities. People can continue to appreciate the mechanistic approaches of behaviorism and science, but they must also cherish logic, wisdom and emotion.

Creative play should not be stuck away in summer camps and quickie courses (or online), but flourish as a part of our daily home and school experiences. Storytelling festivals, filmmaking, music, drawing (with paint and crayons), individual and group brainstorming, sensitivity training and humor draw out our Hunter sides, just as rote memorization draws out our factual Gatherer sides. For an in-depth exploration of these creative and actualizing opportunities, examine Daniel H. Pink's book *A Whole New Mind: Why Right-Brainers Will Rule the Future.*[5] Also, go to www.gocreate.com.

Maybe we need a real good sliming to get the point across that our boys and girls must become more creative. I ran across an advertisement the other day that marketed the best features of a slime-maker. Yes, the Super Slimer does only one thing, but it does it well. It produces material that looks like, yes, slime. To get out of our intellectual cocoon, we need to feel something wet and slimy all over us. Gooey finger paints will do nicely, also.

I've been with Gatherer kids who wanted to paint but who didn't want to get paint on their fingers or clothes, even though the paint was washable. This reminds me of the activity therapy—sometimes called play therapy—I did in my clinical practice. I prefer the term activity therapy to play therapy, because the children's parents are puzzled why a grown man wants to play with their kids and then sends them a bill for services rendered. But you know what? It works.

Some of the young kids I have treated, mostly boys, were tied in emotional knots. They needed to let loose in a safe, controlled environment. They couldn't verbalize their problems, because they were too young to express themselves that way—and they were boys. These kids expressed themselves through spontaneous and sometimes wild activities in the playroom. If my insurers knew what I was up to, they would have increased my medical insurance. With physical, hands-on interaction and a positive role model, the kids blossomed.

The moral of this story is simple: We need to get our bodies up and DO SOMETHING! Our bodies will be grateful.

Goodbye trivial stimulation.

And maybe, if we get real lucky, we'll rebel against superficial narcissism and just grow up. In the article "Goodbye Bland Affluençe" by Peggy Noonan, tells the story of a family that decided to give up credit cards, high-tech toys and all the other distracters for a new kind of future: "pared down, more natural, more stable, less full of enervating overstimulation, of what Walker Percy called the 'trivial magic' of modern times."[6]

Let's hope they represent the future.

17

Suggestions for Inoculating the Family

The primary purpose of this book is to wave a giant red flag and sound the pandemic alarm. Once parents see the potential dangers of the electronic age, they can develop strategies that complement their unique family life styles. This life style varies not only between political and cultural areas around the globe, but also within countries and regions. Unlike medical pandemics, this digital pandemic changes too rapidly to permit permanent solutions. Before work on a new antibiotic is completed, a variant strain of the virus appears. And, unlike the disease model, IT offers some positive attributes which, if carefully monitored, can aid humanity and family cohesion.

What's a parent to do?

Setting up and enforcing parental strategies requires large doses of one powerful antidote: CONFIDENCE. It always amazes me to see how intimidated some parents are because their children know more about an electronic application than they do. While a child is rattling off a slew of new interacting applications, the parent smiles in awe, like some country bumpkin who just got off the bus in New York City. Remember, what you learned about life back in Watertown, Wisconsin, is more important than tall buildings—and people in a hurry.

One useful analogy is the camera. Some folks are technology experts and use all possible camera applications. Does this technical skill make them photographic artists or great photojournalists? Of course not. It's not the technology, it's the person using the technology that counts. And that person is shaped by his or her family, education and values. A few years back, scientists debated how best to harvest accurate and comprehensive data from exploratory field trips. They finally came up with a foolproof method. They decided to require the use of a camera. Instead of a notepad, the technicians took a camera to explore a community in Africa. But when they came back, scientists discovered that important areas had not been documented—and that's because it all depends on where you point the camera!

In World War II, psychologists attempted to identify top guns, the pilots who excelled in combat. A camera, located in the nose of a fighter plane, showed actual combat footage. This would certainly identify the best shooters. Correct? Once again, technology was upended by the human mind. It turned out, unbeknownst to the psychologists, that fighter aircraft flew in certain formations that were defensive as well as offensive, and one designated pilot had a higher probability of a "hit" than other pilots in the formation.

Speaking of underlying human concepts, let's remember what would typically happen back in 300 A.D. when a businessman left Rome to travel to Athens. He'd try to find a good sailing ship that would get him to Athens on time, and once he arrived he had three basic decisions to make: Whether or not to be faithful to his spouse back home, whether to be honest in his business dealings and where to find a good hotel.

Today, it's no different. A businesswoman going from Rome to Athens flies in a jet instead of a sailboat, but faces the same practical and moral decisions. She might carry an iPhone with GPS and apps with restaurant information, but the moral questions remain the same. By the way, Conde Nast Traveler magazine sent three reporters to Moscow in 2009, one with an iPhone, one with a Blackberry and one with an old-fashioned guidebook. The writer equipped with the guidebook completed most of the typical tourist challenges, such as finding a hotel or restaurant, more quickly and easily than the writers with the electronic gadgets.[1] Here's a salute to old-fashioned ingenuity!

The moral of these stories is that parents need to keep an eye on the outcome of their children's forays into the high tech world and not the process or how well the technology works. You have basic goals and values in mind for your children. Is technology helping to promote those goals and values or is it contaminating your child in some way? Don't let the razzle-dazzle of the latest electronic gimmick distract you.

Are you intimidated because your child knows more than you do? Remember, your kids may do many things better than you. They might have greater musical talent, better phonics skills or run faster. Wonderful! You want your children to excel, but this has little to do with maturity and wisdom. Just remember: You're the parent. A teen daughter knows more about applying makeup than her father, but he knows how it makes her look and how it impacts her friends.

And one final point. Don't fall for the generation-gap argument. It goes like this: These kids are Digital Natives who have grown up with electronics. Their parents are Digital Immigrants who can't keep up with technology. As a result, parents are afraid of something that is perfectly natural for the young folks. Leave the kids alone, and let them lead us into a brave new world. They're right about one thing. Parents are fearful of these unchartered waters—and have every right to be.

Remember the Developmental Principle: Kids aren't midget adults. When parents watch their young child play a basic electronic game, they're not seeing what the child is seeing. The child's brain is less developed and highly impressionable. The child is responding in a more concrete manner than the adults and lacks the ability to hold back or withdraw from the rich, reinforcing stimulus.

This line of reasoning is often used to support "children's liberation." If adults aren't seeing what the kids are seeing, how do they know how the game is affecting their children? Where do adults get off advising their kids when the kids are learning in their own way and are in their own special world? While parents can't experience it the way their children do and probably wouldn't want to, they have the maturity and wisdom to look at outcome factors and draw on the experience of folks who study such matters.

It's my contention that people who didn't grow up with electronics from the age of three (Digital Immigrants) are benefiting more from the electronic age than those who did (Digital Natives). That's because they have experienced a more personalized, humanistic education, family cohesion and deeply embedded friendships. They can put the new gadgets and their benefits into the context of their lives and the history of their times.

Here's one tiny albeit concrete example of what worries me. At some point not too far down the road, five-year-olds will be given the latest cell phone when they graduate from kindergarten. The GPS application will direct them through their schools and neighborhoods. Sounds good. But it also robs their brains of important spatial reasoning and directionality stimulation.

So what happens when the battery runs out or there is a malfunction with the gadget? At age ten, these kids won't have the same brains as kids who developed their knowledge of the environment from the soles of their feet and the changing panorama of light and wind. They will be on the earth but not of the earth. And ten-year-old Digital Immigrants, if any still exist, will use the gadgets even better than their cyber twins. They will understand the visual mapping feedback and respond better to oral directions. They may not even want to use the GPS, because, like Tom Sawyer, they enjoy wondering around freely and taking satisfaction in finding their own ways.

Now apply this example to a whole myriad of emotional, tactile and intellectual skills and the concerns are evident. Does this mean we are losing our kids to the digital pandemic? Should parents and legislators push for government intervention? Not yet. If families put first things first and prevent their kids from becoming habituated to all things electronic, to the detriment of known building blocks and best teaching practices, their children should be fine.

Some government intervention will undoubtedly occur. It already has. Some states have made it unlawful to drive while using a cell phone, just as they have required the use of seat belts. Texting while driving will also become unlawful. In the long run, education is probably the best answer. Tax payers will insist on more research to examine the effects of technology on the brain and central nervous system as well as our physical and mental health. Philosophers and religions will draw on history and wisdom to study its impact on humanity.

All about priorities.

Focusing on digital technology can be overwhelming. New electronic devices are introduced weekly—and children want them all. Yep, keeping up with pestilence-spewing robots and the kids can be stressful. The answer is to start with what's needed for a (generally) happy family and not over-focus on electronics. If we do the important things first, the digital lifestyle can take its proper place within the context of human family activities.

It often helps to draw a simple chart that includes a behavior checklist, chores, free time or "goof off" time, outside play time, sibling time, etc. It could include bedtimes, study times and reading time—and yes, cell phone, computer and electronic game time. Hunter personality-types might prefer a chart that is posted at a center point on the refrigerator or near the dinner table, while Gatherer parents may want to write out these basic rules and guidelines or present them verbally. Bringing back a nightly family dinner time is an anchor that pays many dividends—even if the kids claim they're bored.

The behavior checklist could include items such as helping parents set the table, not provoking siblings, reducing whining, etc. For older kids it could include things like taking out the garbage, putting the bike away, etc. Charts work. They are especially valuable for setting up structure. Later they can often be discarded. The main thing is to start with a reasonable family structure and remain consistent. Then find a place for electronic devices and communications.

How much time should be allocated for use of technology? Parents need to come up with their own guidelines, but here are some suggestions: For children above the age of eight, limit electronic game time to thirty minutes per day during the school week and sixty minutes per day on weekends and in the summer. Of course, this does not include games labeled Mature and excessively violent games. Thirty minutes a day, school and weekends, is enough exposure for younger children.

Keep in mind that if these games just involve shooting and blowing things up, or require dexterity and balance without the need for problem solving, there isn't much going on in the brain. If your child is fulfilling most of the basic family functions as outlined on your family chart, taking a short break from thinking and wasting a little time may be fine.

Your three- to seven-year-olds are most vulnerable. The developing brain benefits from stimulation, and basic games don't provide that needed intellectual excitation. Even though television is often described as a great waste land, some offerings are excellent and most feature real people as actors or nifty cartoons with valuable lessons. With basic electronic games, it isn't the content that is the only concern, it's the lack of thinking and memory required and the amount of time spent. Check with your pediatrician for evolving age-related guidelines on exposure to television and electronic games.

The worse case scenario is a lifetime of responding to outside stimulation without developing curiosity and imagination, like the pug in the park who was stimulus bound by the moving light on the sidewalk. We don't have the research yet, but we're already seeing more of this passivity and dependence in teens. How will they turn out as adults? Just look around.

There are some adults who are not creative and imaginative, yet they stay busy and do well in a structured business environment. When they retire, they don't have a clue about what they want to do. These people, even without exposure to gaming, have reached this passive, dependent state their own—without the digital pandemic. They are probably extreme Gatherers. Now, imagine the extreme Gatherer who is brought up "inside" a computer. That person won't have the skills or temperament to succeed in the labor market or the social and friendship market.

You will eventually see advertisements for computerized brain stimulators. These programs are designed to improve thinking skills and some schools have incorporated them into their curricula.

Should you invest in one of these brain enhancers? Some modest gains have resulted from these programs when used by seniors who have lost memory and visual skills, but I think it's too early to use them extensively with children, especially children who show no signs of brain dysfunction or learning disabilities. I'm not too sure about older folks, either. Developing new interests that tap into unused brain centers seems to help the most. The accountant who takes art lessons is a good example.

Electronic game activity is a privilege, not a basic human right. It is fair and right to use gaming as a reward for completion of other activities. When you see your child playing an electronic game, ask yourself this question: What is he or she being rewarded for? Following Grandma's Rule, "Work before Play," we expect game usage to follow, not precede chores and study. If the child engages in homework and accomplishes certain teacher demands or empties all of the trash baskets as required, then it's on to happy blood and mayhem time!

A family that watches together...

What about the Internet, cell phones and television? Smart parents watch TV with their kids. This can make for lively conversation and is an opportunity to examine popular language and behavior in the light of family values. Parents hoped that successful strategies such as conversation and interpretation would carry over to internet usage, but research to date has showed that only banning and restricting the internet prevents overuse and online risks.

How come? By comparison with television, it is difficult to make internet use a shared activity. This is because of screen size, sitting position, the mouse and the common location in a small or private room. Who has the greatest likelihood for risk? It's boys and those who have greater online skills.

Some parents believe their teen kids can't be contained. "If we restrict violent electronic games, even those labeled Mature and Adult, our kids will just go to their friends' homes and play the games there." This is a cop-out. If parents don't make reasonable efforts to guide their children, this may be interpreted by their children as indifference and a lack of affection. Kids can and sometimes will take advantage of their friends' parents who don't care or are away from the house, but they expect their own parents to be appropriately vigilant and caring. Maybe they can take drugs and watch X-rated movies at their friends' homes, too, and ride with friends who text while they drive. Does that make it okay? Does that mean parents should give up?

Technology does allow for filters, monitoring and checking on the child's activities, overtly or covertly. These approaches raise problems, however. Parents need to be the boss, but this has to be weighed against the child's normal desire for autonomy and independence. If you believe you need to monitor your child's driving through the use of onboard cameras or to examine your child's Internet messages, sit down with your child and discuss your concerns before implementing these systems. If your child is able to respond to the basic family structure and values, this approach may not be necessary.

It's probably wise to limit Internet access to times when a parent is at home and can monitor what part of the world—or underworld—the child is visiting that day. This can be accomplished by using an entry code for computer use that is held in TOP SECRET by the "nasty," "authoritative" parents. The computer is best positioned where parents can observe usage, at least until the age of sixteen or so. The computer may become part of schoolwork production (notice that I didn't say time). But some of the homework should require the child's head to come away from the computer screen and into a book. Note taking and writing are also irreplaceable.

The cell phone may function as a social umbilical cord, but should be separated from the child during meals and other formal family functions. Depositing the cell phone in a designated box at certain times of the day and at bedtime is important for a basic sense of structure. Having the child converse with friends for hours at a time during the night is not conducive to academic excellence or good health, and the cell phone vibration feature allows easy access to this kind of clandestine activity. It won't be long before most computer functions will be embedded in cell phones and this will add even more challenges to family cohesion and children's egos.

Preventing malnutrition.

When you come up with your own unique family structure and activities, keep in mind the concerns highlighted in this book. This will help you introduce "curative ointments" to counteract the digital pandemic. Here's what the great plague has wrought:

- Poor directed attention and concentration.
- Low frustration tolerance.
- No time for the joy of anticipation.
- Little solitude. (Old Turkish saying: "Only God is entitled to solitude.")
- Poor working memory.
- Limited development of imagination and creativity.
- Poor communication skills.
- Shallow online friendships. Let's call them relationships, not friendships.
- Inability to prioritize because of attempts to "multitask."
- Inability to persist for long periods of time at one activity such as reading.

Encourage your children to lengthen their attention spans and working memories. These can be reinforced through playing games that require some level of intellectual skill such as cards and board games; and homework, of course. Children need to learn how to tolerate frustration and boredom. Okay, sitting in a restaurant waiting for table service can be boring, especially for kids, but rather than whipping out handheld video games and other electronic devices, your child might have to invent something creative with a napkin, actually look at people and restaurant furnishings or even—heaven forbid!—talk to siblings and parents. Why do we think it's odd if a child reads a book while waiting for food in a restaurant but believe it's perfectly natural to play a computer game?

Your kids need to play together, even if it leads to fighting and squabbling. Having them sit inside, in stony silence, interacting with a machine may be good for your nerves, but won't help them learn lessons about cooperation, frustration and fair play.

Develop persistence by assigning tasks to be completed by work produced, not time spent on the job. This could apply equally to chores such as yard work or homework. Give your child a short book to read, all at one time, and with no interruptions—that means no TV, computer or cell phone. Let him enjoy the self-affirming rewards of persistence...and solitude.

Is reading the magic pill that heads off the digital pestilence? If so, why not buy your child an e-book? With this device your child can access thousands of books without going to the library and without waiting for a new book to arrive. (We wouldn't want them to develop frustration tolerance and the joy of anticipation, would we?) I think e-books are fine for Digital Immigrants who already know how to "get into a book and stay there," but our Digital Natives have been conditioned to rapid change and low levels of persistence, not total immersion in reading.

Unfortunately, e-books can foster that predisposition. Because of quick accessibility, the reader is tempted to change books on the spur of the moment. They can now be added to cell phones and the result will be lots of distractions. Writer Steven Johnson fears that e-books may not only increase distractions, but also lead to the end of reading alone. Readers can take passages out of books and comment on them in public. Indexing and ranking individual pages—and even paragraphs—based on online chatter is becoming a reality. Johnson believes we may start reading books like we read magazines and newspapers, "a little here and a little there."[2] Again, digital immigrants can make good use of this approach, but our kids? I'm not so sure.

When I was an undergraduate, in medieval days, my college turned the lights off at 11:00 P.M. each weekday night. No, this wasn't a monastery. As I look back now, I realize how much this helped me to organize, plan and develop frustration tolerance. I also learned to appreciate the value of flashlights and candles!

The joy of anticipation comes from waiting for something. If your child orders an orange and black jacket on the Internet—with your approval, of course—use regular shipping, not expedited shipping. If your child wants to go shopping because she just "has to have" some cool clothing item, plan the trip for a few days hence. Try to slow things down at home and follow your own charted guidelines. I recall waiting forever for my Captain Midnight

Decoder Ring. I thought it would never come. When the mysterious gold ring finally arrived, it had already turned green. I was never more excited or happy.

Kids need these anchors—now more than ever—to develop solid self concepts. Their brains and bodies are constantly changing, just like the apps on cell phones. And when we add too many electronic games and communications—not to mention avatars—we may be asking for trouble. Children develop their self-concepts through "I, me and mine" experiences. Much of the valuable feedback they need comes from interacting with real people.

How do we define friendship? Allan Bloom, quoted in the *Wall Street Journal*, states: "Lack of profound contact with other human beings [is] the disease of our time."[3] Writer Tony Woodlief correctly surmises that friendship isn't tangential or voyeuristic.[4] That's true, but as stated earlier in the book, virtual boyfriends and girlfriends are all the rage in Japan. With the advent of Facebook, romantic avatars and other electronic experiences that masquerade as friendship, what's a parent to do? Remember, keep your eye on your basic family structure and give your child plenty of opportunities to share face-to-face activities with age-appropriate children.

As we've seen, MySpace and Facebook are not likely to help. Popular kids may expand their friendship base on these sites—even find more real friends—but the sites don't help the shy child make the contacts necessary for true friendship. These social sites may help the shy child feel a little less lonely, but parents need to step in and help the child socialize. If it's the child's personality, speaking with the child's teacher and school counselor usually helps provide valuable input.

Communication skills are important to your child's ability to make friends. Kids text because their teen culture demands it. What's new about the teen culture and the pressures it brings to bear? When I was fourteen, silky orange and black jackets were in. I've never felt as kingly as the day I donned my magnificent jacket—a comforting symbol of allegiance to my "misunderstood" peer group, other brave kids who were just as confused about life as I was. But teen pressures concern every parent. The Internet opens access to illegal drugs and other temptations. Teens can organize meetings on the spot, thanks to cell phones, but those same cell phones let parents know where their kids are.

It goes without saying that e-cigarettes are out. It doesn't matter if the smoke is free of addictive chemicals that cause disease, as their manufacturers claim. They still have the appearances of cigarette smoking and this will make it attractive to kids who need to be perceived as cool. We don't have the research yet, but I think the behavior associated with e-cigarettes will shape the child's motivation in the direction of smoking tobacco.

Texting and social sites.

There's nothing inherently wrong with texting, as long as it's balanced by the development of proper verbal, writing and spelling skills. Texting is an extremely quick and efficient way to convey facts, but it falls short as a way of relating closely to others. And if kids text too much, their brains cells will change to reinforce this skill. That's fine, but what about proper writing, in the country's lexicon, which is necessary for raising a family and putting food on the table? Use it or lose it!

Your kids need to be warned about sending pictures to social sites that can cause them embarrassment and even legal problems. Do they realize that everything they give to the Web site may be owned and controlled by strangers? While you're at it, warn them about online bullying. You don't expect your child to bully others, but this isn't similar to physically abusing a smaller child on the playground. This is probably much more serious, because now we're playing with people's minds...and hearts. It's easy to slip from sarcasm to heckling to harassment to stalking. Most serious crime has its etiology in pride, envy and loss of self-esteem. If your child can't get along with someone, it is better for your child to let go and break it off. In the old, old, days it was dueling pistols at dawn; now it's digital damnation.

What are the best ways to communicate? I'd put face-to-face verbal communication with a good friend in the top spot. (It's nice when this includes spouses as well.) This involves verbal and nonverbal skills. Our past association with our friends sharpens our perception and understanding and kindles trust and self disclosure. Add physical touch to this form of communication—a handshake or a pat on the back—and we have all the necessary ingredients for good communication and a continuing friendship. Communication is like love: It deepens over time. Texting is just a road sign, while true communication is a joyful and enlightened spiritual journey.

What comes next? I suppose talking with friends on the phone is second. We can't see them, but the unique rhythm and intonation of their speech draws us in. I'd put writing a letter in the third spot, followed by smoke signals, E.S.P. and finally, texting. Just kidding.

Parent-child communication is an important key to family cohesion—now and in the future. This is direct, face-to-face communication, not texting. Girls and Gatherers are usually more verbal then Hunters and boys, but communications can be non-verbal as well as verbal. Wrestling, tussling and tickling are great ways to communicate, especially at younger ages. Hugging can go a long way. Apparently, teachers and school authorities are observing a lot of student hugging, to almost an obsessive degree. This could be a reaction to feelings of loneliness and superficiality wrought by Gaderian, the pandemic robot and gatherer of kids.

One way to draw your child out verbally is to "listen with the third ear." Instead of focusing on the content of your child's speech, listen to the emotion conveyed. Psychologist Carl Rogers used this approach almost exclusively in counseling. Your teen blurts out, "Mrs. Riley didn't even listen when I handed the paper in—just 'cause it was ten minutes late." You're tempted to blurt out your response and remind your child about punctuality and the stacks of papers Mrs. Riley faces each day. That would be an okay response, and your child might listen, but you could also respond this way: "I guess you're really ticked off. I guess you don't think Mrs. Riley was being fair." This might keep the conversation going for awhile. There will still be time to deliver the valuable message about timeliness and responsibility.

Here's a suggestion that might come as a surprise to some parents. Teach your child computer skills. Wait a minute. Aren't these Digital Natives expert in everything electronic? As reported in chapter 7, our kids are so mesmerized with electronic games that they're losing interest in computer programs. If the school can't help your kids keep up, you may need to instruct them at home. The public library can help.

Public libraries can help in many other ways, as well. In fact, they represent one of the antidotes that really work. They incorporate many of the values mentioned earlier: Quiet, solitude and germ free books everywhere. Talk about a bargain; the price is right and the staff is well trained. Recently some politicians illustrated their opposition to proposed legislation for a government-run insurance program, warning that it would operative inefficiently. They compared it to the service one receives from the U.S postal service and county vehicle registration offices. Interestingly enough, I never hear anyone claim that we have unfriendly or inefficient librarians. They must have excellent hiring and training standards.

Our public libraries are well run and offer a palliative environment that can combat "techie-nitis"—a word I just invented. It means undue inflammation of the technical lifestyle. Here's a place where children and families can learn to read and develop their creative sides. In St. Petersburg, Florida, where I live, the *St. Petersburg Times* and the Humanity Council sponsor a library program called Prime Time. It's a six week program that teaches parents to spend quality time reading with their children.

According to Elaine Birkinshaw, manager of the main library in St. Petersburg, Florida, they use puppets, magic shows and other magnets to bring in children. Volunteers from the Junior League have put on puppet shows for the past twenty-five years. When families are unable to visit the library, her staff goes to schools and summer day camps to introduce kids to the world of reading.[5] Some Digital Natives think the library is a dinosaur that has no place in today's high-tech world. I think it's a lifeline that we should support and appreciate.

We need our finest, socially-conscious companies to reduce the harmful by-products of technology. Let's not be naïve; companies and their stockholders want a good financial return for the risks they take. Can tech companies develop products that foster child and family development and still make money? Does the machine culture have to be the antithesis of humanity?

If companies are able to cut down on pollutants and generate clean air, perhaps they can also help maintain the balance between technical efficiency and humanity. Perhaps it's in their best interests —even considering the bottom line—to head off excessive government intervention and lawsuits growing out of addiction, brain dysfunction and automotive fatalities. Tobacco companies warn youth not to smoke. Perhaps tech companies can help the consumer draw out the best in technology and sidestep potentially serious problems.

Conclusion

Mechanization and the Hunter Spirit

The encroachment of mechanization, I believe, is destroying the human spirit. We have seen the relentless movement inward from Hunter and Gatherer to agricultural farmer to industrial worker to information economy to office automaton to post-modern thinker to cut and paste modern browser. Are we moving from active, sensitive and creative to passive, mechanical and rigid?

Not only do psychological studies raise concerns over the "machine's" negative impact on individual learning and development, but the growth in government regulations, legalistic thinking, objective testing in schools, increasing dependence on electronic gaming and the slicing and dicing of humans into abstract categories all point to the power of machines to influence our minds and destroy our souls.

Gatherers and Hunters.

Mechanization poses the greatest immediate threat to the Gatherer because technology flatters the Gatherer's logical and conventional style. The Hunter's active and intuitive style is also attracted to electronic games such as first-person shooters and other adventurous formats.

When we look at the extreme levels of the Gatherer and Hunter personalities, we realize just how different they are. They think differently, they listen differently, they speak differently and they don't get along very well. They seem to ignore each other because their differences make communication difficult.

The Gatherer speaks and thinks in a detailed manner from the bottom up, while the Hunter thinks and speaks with a large brush, from the top down. Of course we probably all have aspects of these personalities within us, to some degree, and sometimes we can feel a tug of war within ourselves.

Family cohesion.

This is the information technology movement's greatest threat to humanity. In discussing family-breakers twenty years ago, there was a focus on parental over-involvement in work, social and civic activities, and children's emersion into the peer culture. Of course, the television set was a factor, but usually the family watched television together, at least part of the time. This permitted an exchange of ideas, some control over harmful content and even conversation. Now, the children may be watching television or playing electronic games alone in their bedrooms.

When asked a question by parents, many children look up from the dinner table with a vacant stare and mumble, "Huh?" Most are occupied during the entire meal texting a friend or two (or three).

How much time did teens want to spend with their friends in past days? I'd guess 90 percent or higher. Why not 150 percent—remember your teen years? But because of limited telephone time and transportation problems, teens were forced to put up with their parents and even their siblings. Peer contact was reduced to a mere 50 percent or so.

Today, kids haven't changed, but the hardware has. Teens still crave 90 percent peer contact, but now instant messaging and other IT facilitators have, in fact, given them what they want—close to 90 percent engagement. With more women working and parents massaging their beloved computers, shared family time is harder to find. This cohesion, based on and strengthened by respect and emotional attachment, can't be replaced without significantly undercutting the individual, the family and society.

Privacy and dignity.

The IT movement, aided by legal and scientific intrusions, is manipulating us into abstract categories such as age, gender and job title. When we are viewed in this abstract way as animals or statistical entities, there is little need for privacy or dignity. What about the new security machines that strip us bare at airports? The machine is at it again. Yes, because of our machines we're pushing the envelope when we refer to human dignity. Could that old-fashioned dignity idea become a thing of the past? What about privacy? Those with cell phones can take your picture or transmit your actions to people in other parts of the world in nanoseconds.

There's been some debate about the hoards of data shared by 200 million users of Facebook by a consumer advocacy blog, Consumerist.com, that posted an item with the following headline: (later retracted after the impact hit):

"Facebook's new terms of service:
We can do anything we want with your content.
Forever."[1]

Solitude.

Technology inundates us with stimulation from morning to night (and sometimes *during* the night). It demands our constant attention and pressures us to think about work, work and more work. Driving in our cars or taking thirty minute lunch breaks used to give us some solitude. No more.

Philosopher Jacques Ellul asks, "What does technology allow us to ignore?"[2] Christine Rosen, in reporting on text-to-speech technologies, states, "They also allow us to ignore something else, something potentially more valuable than the supposed efficiency we gain by using them: the opportunity to think, to let our minds wander away from work and mundane concerns. They let us ignore the possibility of silence."[3]

Trust.

As indicated in chapter 1, we have seen a progression from a knight's open visor to a military salute to a handshake to today's digital reading of veins in our hands. Today, the personal, face-to-face assurances of trust are replaced by cold metallic hardware. Our veins may be warm, but our blood is turning cold. Uncle Gaderian, the pandemic robot, rides again. This "welcome" takes place in only seconds, and once we're in the system, we are in the system forever.

Electronic Games.
It's important to realize that basic computer games are quite different from advanced games played by older teens and adults. In chapter 7 it was noted that learning usually proceeds from the concrete to more abstract levels in school subjects and even in intelligence testing. With these games there's no red flag to alert us to the fact that we've just crossed a threshold into another more abstract category of learning. Basic games, such as single-shooter and racing games, fall into the concrete category.

When parents look at these games, they focus on the content. Looks harmless, doesn't it? But it's the *process*, not the *content* that poses a risk to the developing brain. Parents won't usually find excessive violence and mayhem at the basic levels, although they will undoubtedly come. What they will find is a mind-numbing, repetitious activity that doesn't stimulate important areas of the brain and doesn't exercise the child's working memory. These games *do* stimulate unimportant brain cells that will not help children comprehend the broader world around them. Gatherers are attracted to the mechanical aspects of hardware at this level, including computers, cell phones and game consoles.

However, my experience and research has indicated that playing more than an hour or two hours a day probably impairs the child's visualization skills and attention span. It's too early to determine the full impact of this bombardment of the young, vulnerable brain. At best, these basic games are a waste of time, and they're competing with more important activities such as spending time with siblings and parents or playing such games as cards, checkers and chess (with humans) that stimulate important areas of the brain. Too much gaming even impairs computer skills, so now we have one IT machine battling another IT machine.

They're low-level games, pure and simple, but many feel they do a great job of keeping children quiet and out of trouble. Perhaps they should be viewed as on-call babysitters—like magical fairy godmothers—who appear at the touch of a button. They're mighty convenient, but beware the epidemic implications. Small contacts probably don't infect the individual's intellectual development, and their use can be employed as a reward for accomplishing more important things, such as homework or chatting with Grandma.

Advanced electronic games and massive multi-player online role-playing games are designed with gorgeous and seductive graphics, require multi-level problem-solving, exercise working memory and do teach *something*. But, what do they teach, and why and how well...and how long does it take?

Again, these games were designed for entertainment purposes, not as IQ builders. Passing them off as educational opens them to the enormously lucrative market generated by our public, *tax-supported* schools. That means we'd pay the bill. Here are some other potentially infectious signs (manifestations) of so-called advanced games:

- Like with the basic, more concrete games, they compete with important activities such as schoolwork and family cohesion. Hunters love the aggression and violence found in some of these games.
- Even though their content is not appropriate for children under the age of sixteen (or maybe any age, for that matter) a high percentage of younger kids play them.
- They don't conform to *Best Learning Practices* and are not designed to teach important educational and intellectual skills.
- Learning density can be defined as the amount of learning over a given period of time. This ratio can't be good, because these games take a long time to teach whatever it is they teach.
- Excessive use of violent games increases aggressive thinking and behavior. As with the basic, more concrete games, excessive use leads to inattention, poor creative visualization and lower grades in school.
- Games at this higher level are more addictive than basic games because they draw on intrinsic motivation.
- At least these interactive games are better than blankly watching television. I have my doubts about that. For one thing, television shows are about humans, not pulsating comic book characters. Even the dumbest family or office comedy show or soap opera makes us think about relationships, and television offers movies, documentaries and the gut-wrenching drama of college basketball's March Madness. And you can find humor on television as well. Robots can't stand humor—it's one thing, among many others, they can't do.

In his 1964 book, *The Technological Society*, philosopher Jacques Ellul provides "76 Reasonable Questions to Ask about Any Technology." Here are some questions that raise concern and need further scrutiny and longitudinal research:

What are its effects on relationships?
Does it undermine traditional forms of community?
Does it build on, or contribute to, the renewal of traditional forms of knowledge?
To what extent does it redefine reality?
Does it erase a sense of time and history?
What is its potential to become addictive?
What values do its uses foster?
What are its effects beyond its utility to the individual?
What is lost in using it?
What are its effects on the least advantaged in society?
What does it allow us to ignore?
What behavior might it make possible in the future?
Does it alter our sense of time and relationships in ways conducive to nihilism?
What is its impact on crafts?
Does it reduce, deaden or enhance human creativity?
Does it depress or enhance the meaning of work?
Does it express love?
What aspect of our past does it reflect?
Does it reflect cyclical or linear thinking?
Does it undermine traditional moral authority?
What noise does it make?
What pace does it set?
How does it affect the quality of life (as distinct from the standard of living)?[4]

I hope I have been an informative pandemic predictor. If you know when the epidemics are coming, you have time to protect yourself and your family by taking proper precautions, limiting exposure and hopefully creating vaccines. If we use our machines rather than letting them destroy our humanity, we'll live in a productive, creative land for many years to come. In other words, we'll all live happily ever after.

Appendix

Feed a Cold, Starve a Fever:

Suggestions for Schools

An ounce of prevention.

Let's start with the basics: The body. It's easy to forget the importance of our physical makeup and the critical need for physical activity. This is especially true in the classroom, where the bottom line is clear: "Do well on the big government-mandated test at the end of the year—or else." Some folks believe it would be much simpler just to forget about recess and let filmstrips, television and computers give us a view to the outside world. Isn't it easier to control students in a structured classroom than outside? Absolutely. But what a price we pay.

Perhaps it's the price the Hunter pays. As discussed earlier, the Hunter personality, especially the male Hunter, is drawn to spatial and kinesthetic activities. These kids are desperate for recess. It not only conditions the body, but clears the mind, reduces attention problems and aids in the development of the brain. And we sometimes overlook the fact that recess provides Gatherer types, especially females, an unstructured socialization time. It often lets Hunter girls and boys, who enjoy and excel at sports, earn the respect of their classmates. With the mass inhalation of pandemic fumes from junk food, our kids need exercise even more than their parents and grandparents.

The theory of Gatherer and Hunter personalities has hopefully given us better insight into how different people respond to the digital pandemic. If teachers can see some of these personality characteristics in their students, it may help them as teachers. The Gatherer type feels at home with computerized instruction but doesn't need it to improve motivation. This student is usually happy with logical and sequential learning—much like a computer—and has a good memory for details and facts. He or she is usually verbal and is a good test-taker.

The Hunter child has some problems with sequential learning and may not like phonics instruction. Since he or she is highly visual, instructional computer programs may help with motivation and learning. These kids are often adventure seekers and risk takers. They are drawn to violent video games, but also crave movement and need outdoor time, preferably on green grass.

The concept of multiple intelligences has been around for some time. Whether or not these identified skills and traits can be called separate intelligences is beyond the scope of this book. What I have identified as the Gatherer personality would encompass so-called linguistic intelligence, which emphasizes people who are highly verbal and have a strong ability to retain facts. It also includes the logical mathematical mind where there is an emphasis on logic and numbers such as seen in the scientist, accountant and computer programmer. Intrapersonal intelligence also falls here. Introspection and contemplation are important attributes of this type.

The Hunter personality encompasses spatial intelligence which involves thinking in pictures and images. Many of these people enter the visual-spatial vocational world as architects, artists and pilots. It also includes so-called bodily-kinesthetic intelligence. These folks become athletes, craftspeople and mechanics. Silent screen star Charlie Chaplin, who could communicate complex ideas entirely without speech, was the epitome of the Hunter personality. They also fall in the interpersonal intelligence bracket. They make great sales people, social directors and negotiators.

What about structured exercises, however? Physical education researchers Grant Hill and Bud Turner came up with a checklist to promote fitness and physical activity. Here are a few items from that list:

- Posters of males and females exercising are visible to students.
- Team size for traditional field sports such as softball, football and soccer is kept at 7 members or less.
- Elimination games are either not used or the rules are altered to keep all children active."[1]

These alterations to standardized activities are helpful, because of the limited time available for recess in some schools. Their main goal is to keep all children active. At the same time, games need to be fun. Athletic games without fun are just exercise—yuk.

Basic epidemiology.

My suggestions for school boards, schools and teachers begin with the same advice I give to parents: Start with a good basic classroom structure, have confidence and THEN worry and wonder about technology. If you're a teacher reading this, you're probably thinking, *Spare me—more advice from another expert who has never set foot inside a classroom.* Actually, I have set foot in a few hundred classrooms, and I've had to duck a couple of times in some of those free-for-all war zones. But I've never had the responsibility of a head classroom teacher. I can't tell a teacher how to run his or her class, but maybe we can all benefit from sharing ideas about the digital pandemic.

I think most teachers agree that the classroom environment is similar to Maslow's hierarchy of needs, which were cited earlier. We begin with safety and security and then work upward to basic academic skills and advanced academic and thinking skills, using best teaching practices. Incorporated into this hierarchy of educational needs is self-actualization and personal development which include art, music, group cohesion, interpersonal communications, outdoor free play and physical education, to name just a few. None of this works without a good, hands-on teacher.

Even establishing safety and security can be difficult in some schools. School boards fear lawsuits and sometimes back away from appropriate consequences for misbehavior. Students who disrupt classrooms on a regular basis must be moved to special classes where they can receive help for their behavioral problems. And if the school principal doesn't support the teacher, all is lost.

Where does technology come in? Some people think following Best Practices in teaching isn't necessary and can be replaced with computer instruction and electronic games. Technology can help teachers create lesson plans and stay in touch with students and parents. Again, this is a double-edged sword. Students who share their private lives with teachers on MySpace or Facebook can create ethical problems for themselves and their teachers when they report activities that compromise them or their schools. This information snacking also lowers embedded communication to the level of fact sharing—not wisdom and knowledge.

In chapter 5, we acknowledged that corporate shareholders encourage the production of irresistible electronic games for profit. In the school environment we have to the hold politicians responsible for pushing computers and electronic education into the nation's public schools too quickly. Of course, legislators are lobbied by companies to earmark money for electronic systems. Over the years, I've witnessed our schools introduce one educational fad after another, only to discard them. Electronic entertainment and technology could well be our next great educational folly.

What about in the classroom? Computers in the classroom tempt teachers in the same way electronics tempt the parent at home. They can fulfill a babysitting function. But unlike electronic games, they can sometimes help students' break through an impasse that's holding them back. This is especially true for Hunter boys who are strong visually and who enjoy the excitement and action of an instructional game. How does a math fractions instructional game in the classroom differ from an entertainment electronic game? One is instructional and one is entertainment and never the twain shall meet—at least not so far.

They are cousins, however, and we need to be vigilant or we could end up with Gaderian, the pandemic robot, teaching our classes. This is especially true for younger children who are easily shaped into the mechanical world. While computer programs do activate the frontal lobe and working memory, they also take time from other more global types of learning. Kids need to engage in verbal interaction, movement and touch. They need to act on the environment, not just passively react to a mechanical screen, if they are to develop an integrated brain.

Speaking of mechanical screens, it's important to know how much screen-time—as opposed to face time—a student is getting. If we add up television, instructional computer work and instructional movies at school with electronic games, Internet, cell phone and television at home, we've opened the door to a pandemic plague that can sweep through the brain. Oh, and let's add e-books to this list, because the Center for Disease Control is on the way.

Metallic Shakespeare.

What could be any more useful than electronic book readers? Proponents of e-texts believe these glass and metal devices can cut costs and improve learning. There's nothing heavier than a bag full of books, and e-text search functions take the reader directly to the desired information. No paper involved, so we'd be saving our mighty oaks as well. My, but these slick plastic books are handsome and beguiling. I won't repeat the concerns expressed in chapter 15, but most of them apply to the classroom as well.

Ryan Knutson and Geoffrey A. Fowler report that educators are concerned, because digital readers have expensive starting prices and are difficult to share and print. Highlighting, note-taking and sharing capabilities may be lacking. They report on a number of e-book trials, and the results have been mixed.[2]

If anyone draws the line on the electronic invasion, it will be the classroom teacher. Yet some teachers believe kids are not achieving because our classrooms have not kept up with the digital revolution. "They can't learn our way, so we've got to change," they say. Please listen. Your way, not their way, is the only way. Or, put more correctly, your ways are the only ways. They represent a whole array of strategies and instructional methods that are consistent with brain maturation and social and emotional development.

Technology can become another tool that assists in the education process, but it must be tailored to what we have learned about the brain and what we know about a rounded education. Those who get carried away with the romantic idea that the technology will usher in a totally new way of learning better prepare for heartbreak, much like teens who swoon over the latest vampire craze. There is no substitute for a warm-blooded teacher.

If the digital pandemic has already infected some kids, the solution isn't to make the kids sicker. Unfortunately, we've already tried that strategy in other areas, and it hasn't worked. Digital maladies aren't the only ones that produce inattention and lack of motivation. When poor and minority children came to school unmotivated to learn, educators placed the blame on their home environments and, in some cases, gave those kids easy work and lowered expectations for success. "If they drop out it isn't our fault," they said. Digital malnutrition is similar to organic malnutrition.

Maybe we need a digital "free lunch program" for undernourished kids to help them cope with technology-induced deficits. We already have private summer camps for obese kids and those addicted to electronic games and the Internet. Maybe it's time to move some of these programs into our classrooms. Children can be taught to visualize and use their imaginations, for example, and it's critical to teach parents, students and teachers how to use technology in a responsible way—before technology uses us.

The parents' dilemma.

Parents are in the same bind as teachers when it comes to educating their children. Most parents want a well-balanced curriculum for their children, with room for factual, scientific approaches along with experiencing humanity and creativity. Objective testing pushes us toward objective thinking at the expense of subjective thinking. There is a natural tendency for the teacher to teach to the all powerful annual assessment. Even if the parent can find a school that maintains the desired balance, admission to college looms ahead. And this is the bind.

If the child doesn't do well on objective, computer-scored tests, there goes the scholarship or even admission to college. Now we're talking real money and the opportunities a college education provides. Many private schools push even harder for factual results because the reputation of their schools hinges on college acceptance rates while the public school teacher has no choice. Standardized, objective tests are required by law.

So what's a parent to do? Supplementing the formal school experience at home and in the community through the public library and other community opportunities can help, at least until the educational pendulum swings back the other way. This often requires money, transportation and two dedicated parents, however. Introducing electronic education into the classroom at this time is not the answer. It can lead to even less communication and less face-to-face instruction.

Here are some suggestions for choosing a school. If you find one that fulfills all these criteria, you are most fortunate:

1. Smaller schools provide more community spirit. If the elementary principal doesn't know the names of the children and families in the school, it's probably too large. High schools can be larger. The school should value childhood and try to preserve it. This is the Developmental Principle, which was quoted early and often in this book.

2. I believe it's important to hold off the adolescent culture as long as possible. It's better to have grades 1-8 and 9-12 than 1-5, 6-8 and 9-12. Dances for sixth graders who wear stiletto heels and caked, three inch eyelashes are out. Better to spend the evening getting infected by electronic games.

3. The basic classroom structure described earlier should be in place. Computer instruction should be carefully integrated into the students' individualized lesson plans, not used for babysitting or because everyone else is doing it.

4. Well organized classrooms where it is emotionally safe and where students wear modest, non-distracting clothing.

5. No cell phones allowed in school and no electronic games played, even at lunch time, recess and P.E.

6. Nutritional food, packed at home or served at school. No sugar-loaded beverage machines.

7. Art, music and outdoor free play as well as physical education.

8. A good teacher. Someone who loves kids but who doesn't try to be one and maintains control in the classroom. If you have this one, your child will grow, even if you don't have some of the others listed above. If you don't have this one, the rest may not matter.

Newly developed antidotes.

In August 2009, I communicated with Patricia Lambert, Principal of the Center Academy School in Pinellas Park, Florida, about specific measures for combating "techno-contamination."[3] Here are the results of my findings in reference to Ms. Lambert's program:

1. In deference to parents whose children walk to school, use public transportation or ride a bike to and from school and who want the techno-benefits of communication with their children in the event of an emergency, the school allows students to bring cell phones and other technology to school. Even parents who drive their children to school sometimes plead for MP3 players in order to keep their kids quiet in the car. Hopefully, that's not to keep their children busy while Mom or Dad put everyone in danger by cell phoning or texting while driving.

However, students MUST turn in all technology with the exception of CD players at the beginning of the first period of school. Some of our students with learning and attention problems are permitted to listen to instrumental music—no words allowed—when doing selected independent seat work. Lyrics can be distracting at best and offensive at worst. All techno-gadgets are returned to the students at the end of the school day.

2. Internet access is limited and closely supervised. Online gaming is not permitted in school, nor is discussion of online gaming permitted during class time. Online gaming discussions are discouraged during lunchtimes as well, although that isn't always successful. Online gaming clubs are not sanctioned.

3. Students will sometimes express concern over their virtual lives. Others will appear anxious and/or depressed. One student was missing sleep because his online "wife" was contacting him several times a day from a time zone three hours earlier. He and his "wife" had purchased an avatar condo and furniture online. This student desperately wanted a "divorce," but couldn't muster the courage to confront someone he'd never met even once in person! Once his parents learned of his predicament they promptly instituted the world's first online annulment.

When an opening occurs, for whatever reason, it's best to reflect the student's feelings and show compassion for this "airborne illness." Students are led to question the amount of time and money they have devoted to their new avatar lives. Parents need to be brought in to the picture if the child is showing significant emotional upheaval.

4. Parental intervention is necessary when signs of addiction occur. Again, it's important for the school and parents to work together to determine the extent of online and game involvement. It's the total amount of time spent on all forms of electronic exposure, including gaming at friends' homes, that is necessary for a proper "diagnosis." Parents are cautioned to check backpacks before students leave the house for school and remove all books, notes and character implements that are part of the game. Parents usually request that the school discover the materials and confiscate them.

5. Moderation is a word that is missing in the vocabulary of many of our technocrats-in-training. The school needs to teach time management skills to all students. Online activities and total tech time should be incorporated into the students' weekly time-budget plan. Graphs are particularly useful here.

6. Technology exposes the student to information snacking and brief units of action. This immersion into detail puts "big picture" thinking and imagination at risk. This is especially true for the Gatherer who feels most comfortable with detailed and sequential learning.

The use of graphic organizers and chapter surveys is helpful for many students. Techniques used to help reading disabled kids may also aid in the remedial process. One of these approaches asks the student to page through a book and look at illustrations, then ask questions about the pictures and try to anticipate what the story is about before ever starting to read the first word.

7. Again addressing the episodic nature of electronic communications, students have difficulty generalizing from one subject to another. "One of the problems inherent in being 'overtaken' by the digital age is a loss of the sense of time and place," says Lambert. To deal with this, the school incorporates "connectors" between interdisciplinary lessons. A special Friday Lab is an excellent place to practice these connections in order to improve abstract thinking.

The organizer can be any size. It is a paper with the name of the individual or topic at the top just above a row of boxes with topics such as "Education," "Politics," "Personal Information," "Art and Music" "Science and Math," "Social Issues," etc. Beneath each topic is a column of boxes where the student can write in specific information. For example when students read Greek Drama in Literature class, their history teacher refers to the texts they read and relates actual historical accounts, the art teacher shows and discusses architecture, ceramics, etc. and the music teacher contributes music and instruments from that period.

A project on Ben Franklin included history, art, furniture, politics, social issues, music and dance. Dante's Inferno intrigued the students, especially when it came to identifying political figures Dante depicted the fiery inferno. These exercises are fun to do, and the students are able to maintain good levels of concentration.

8. As pointed out in chapter 7, visualization is a problem for many of our IT students. Center Academy teachers working with sixth and seventh graders read short stories to the students and then stop periodically and ask the students to draw what they've heard. "The focus here is usually 'setting' and the exercise enables them to develop skill in visualizing place as well as exercising their imagination. It also extends focus for more than two minutes!" says Lambert.

9. This same approach is also effective with eighth and ninth grade students but with these students the exercise is extended to include atmosphere as well as setting. A favorite tool for this exercise is a collection of Jack London books.

10. High school students are asked not only to visualize setting and atmosphere, but also to extend this activity to character. The art teacher can be helpful here. Students are encouraged to analyze the text for descriptive phrases. This will give them a more exact perception of the character. Discussion and the adding of detail help to convey the character's personality. This bit of "antitoxin" not only extends focus and imagination, but also improves interest in critical, analytic reading.

Acknowledgments

I want to thank my family, friends and colleagues who contributed their insights and experience to this effort, including readers Doug Hicks, Lois Adkins and Andy Hicks. Jim and Sharon Moorhead, *Moorhead Ink*, helped me edit. Andy Hicks, Ph.D., provided additional neuropsychological consultation, and Patricia Lambert offered educational insights. Susan Hicks helped research noteworthy newspaper and magazine articles. Doug Hicks offered ideas on family and business management. Kimberly Hicks conducted teen research. Cecil Cheek and Philip and Joyce Hicks shared their experiences at the collegiate level. I'm indebted to Dan Piraro, Dave Coverly and Rebecca Skelles for creative artistic renderings and Barbara Boucher, Nicki Ryan and Kimberly and Jacob Hicks for indexing references.

The Digital Pandemic would not have been possible without the support and determination of an outstanding literary agent, Sydney Harriet, Ph.D. My deep appreciation goes to Dr. Joan Dunphy, President of New Horizon Press, and her editorial staff for offering their assistance and encouragement in bringing this book to fruition.

Notes

Introduction

[1] L. Gordon Crovitz, "Time to Reinvent the Web," *The Wall Street Journal*, February 9, 2009.

[2] Daniel H. Pink, *A Whole New Mind: Why Right-Brainers Will Rule the Future* (New York: Riverhead Books, 2006).

Chapter 1: How We Got Addicted

[1] Gary Small and Gigi Vorgan, *iBrain: Surviving the Technological Alteration of the Modern Mind* (New York: William Morrow, 2008).

[2] Ibid.

[3] Karen Blumenthal, "Deals Abound, But Which Offer Lasting Delight?" *The Wall Street Journal*, December 3, 2008.

[4] Edward Shih-Tse Wang, et al., "The relationship between leisure satisfaction and life satisfaction of adolescents concerning online games," *Adolescence* 43, March 22, 2008.

[5] Ellen Gamerman, "The New Pranksters," *The Wall Street Journal*, September 12, 2008.

[6] Iwatani Yukari Kane and Danisuke Wakabayashi, "A Way for Gamers to Get a Life," *The Wall Street Journal*, December 17, 2008.

[7] Ibid.

[8] Alex Roth and Paulo Prada, "In Washed-Up Economy, Governor Defends Plan for Anglers' Paradise," *The Wall Street Journal*, January 5, 2009.

[9] F.E. Kuo and Andrea Faber Taylor, "A Potential Natural Treatment for Attention-Deficit/Hyperactivity Disorder: Evidence From a National Study," *American Journal of Public Health* 94, no.9 (September 2004): 1580-1586.

[10] Ibid.

[11] Jimmy Carter, *An Hour Before Daylight* (New York: Simon & Schuster, 2001).

[12] Tom Bradshaw and Bonnie Nichols, "Reading at Risk: A Survey of Literary Reading in America," National Endowment for the Arts *Research Division Report #46*, June 2004.

[13] L.Gordon Crovitz, "The Digital Future of Books," *The Wall Street Journal*, May 19, 2008.
[14] Melinda Beck, "When Your Laptop is a Big Pain in the Neck," *The Wall Street Journal*, December 16, 2008.
[15] Kevin Kelly, *Out of Control* (New York: Perseus Books, 1994).
[16] Douglas Rushkoff, *Playing the Future: How Kids' Culture Can Teach Us to Thrive in an Age of Chaos* (New York: Harper Collins, 1996).
[17] John Palfrey and Urs Gasser, *Born Digital: Understanding the First Generation of Digital Natives* (New York: Basic Books, 2008).
[18] Don Tapscott, *Grown Up Digital: How the Net Generation is Changing Your World* (New York: McGraw-Hill, 2009).
[19] Mizuko Ito, "Digital Youth Project Findings," Pew Internet & American Life Project, September 16, 2008.
[20] Ibid.
[21] Ibid.
[22] Tom Jones, "Michael Phelps Goes Bong," *St. Petersburg Times*, February 4, 2009.
[23] Jessica E. Vascellaro, "Facebook's About-Face on Data," *The Wall Street Journal*, February 19, 2009.
[24] Todd Lewin. "Chips: High-tech aids or tools for Big Brother?" *The Associated Press*, July 23, 2007.
[25] Jason Armagost. "The western cannon," *Harper's Magazine*, September 2008.
[26] William Saletan, "Nowhere to Hide: Killer drones that can see through walls," www.Slate.com http://www.slate.com/id/2200292/?GT1=38001 (accessed September 28, 2008).
[27] Scott Silverstein, Letter to the Editor, *Wall Street Journal*, February 18, 2009.
[28] Siobhan Gorman, August Cole and Yochi Dreazen, "Computer Spies Breach Fighter-Jet Project," *The Wall Street Journal*, April 21, 2009.
[29] *RSA Fellowship: Removing Barriers to Social Progress. News and Events. Fellowship Profile*. HTML Newsletter. www.theRSA.org (Tuesday, 14 July, 2009).
[30] Theodore Millon, "The Renaissance of Personality Assessment and Personality Theory" (*Journal of Personality Assessment*, 1984): 48.
[31] K. Kris Hirst, "Hunter Gatherers," www.About.com http://archaeology.about.com/ od/hterms/g/hunter_gather.htm (accessed July 17, 2009).
[32] Merriam-Webster's Collegiate Dictionary, Tenth ed., s.v. "foraging."

Chapter 2: Halvin and Garrison

[1] "MBTI Basics," The Meyers & Briggs Foundation, http://www.myersbriggs.org/my-mbti-personality-type/mbti-basics/ (accessed July 9, 2009).

[2] Mary T.Russell, M.S., and Darcie Karol, M.A., *The 16PF Fifth Edition Administrator's Manual* (Champaign, IL: Institute for Personality and Ability Testing, 1994).

[3] H.G. Gough, *California Personality Inventory Administrator's Guide.* (Palo Alto, CA: Consulting Psychologists Press, Inc., 1987.)

[4] D. Frank Benson and Evan Zeidel, *The Dual Hemisphere: Specialization in Humans* (New York: Guilford Press, 1985).

[5] "MBTI Basics," *The Meyers & Briggs Foundation,* http:// www.myers-briggs.org/my-mbti-personality-type/mbti-basics/judging-or-perceiving.asp (Accessed July 9, 2009)

[6] Mary T.Russell, M.S., and Darcie Karol, M.A., *The 16PF Fifth Edition Administrator's Manual* (Champaign, IL: Institute for Personality and Ability Testing, 1994).

[7] Aldous Huxley, *Chrome Yellow* (New York: George H. Doran Company, 1921).

[8] John Fowles, *The Ebony Tower* (New York: Vintage Books, 1974).

[9] Joe Morgenstein, "'Hotel for Dogs' is Puppy Chow," *The Wall Street Journal*, January 16, 2009.

[10] Ian McEwan, *Atonement* (New York: Doubleday, 2001).

[11] Thomas Harris, *Red Dragon* (New York: Dell Publishing, 1981).

Chapter 3: Time to Inoculate?

[1] L. Gordon Crovitz, "Lessons from the Great Books Generation," *The Wall Street Journal*, December 8, 2009.

[2] Douglas Rushkoff, *Playing the Future: How Kids' Culture Can Teach Us to Thrive in an Age of Chaos* (New York: Harper Collins, 1996).

[3] Don Tapscott, *Grown Up Digital: How the Net Generation is Changing Your World* (New York: McGraw-Hill, 2009).

[4] Charles Krauthammer, "Man vs. Computer: Still a Match." *Jewish World Review*, November 21, 2003.

[5] Ibid.

[6] L. Gordon Crovitz, "Wikipedia's Old-Fashioned Revolution," *The Wall Street Journal*, April 6, 2009.

[7] W. Bruce Walsh and Nancy Betz, *Tests and Assessments* 2nd Edition (Englewood Cliffs, NJ: Prentice Hall, 1994).

[8] Michael S. Malone, "The Twitter Revolution," *The Wall Street Journal*, April 18, 2009.

[9] Lee Gomes, "Linking Brains, Computers," *The Wall Street Journal*, July 9, 2008.
[10] Sharon Begley, "How Thinking Can Change the Brain," *The Wall Street Journal*, January 19, 2007.
[11] Ibid.
[12] Douglas A. Bernstein and Peggy W. Nash, *Psychology*, 4th Edition (Belmont, California: Wadsworth Publishing, 2006)
[13] Daniel Gilbert, *Stumbling on Happiness* (New York: Vintage, 2007).
[14] Annette H. Zalanowski, "Music Appreciation and Hemisphere Orientation: Visual versus Verbal Involvement," *Journal of Research in Music Education* 38, No. 3 (1990): 197-205.
[15] Dennis Prager, *The Dennis Prager Show*, KRLA 870 AM, October 28, 2008.
[16] Guy Vingerhoets, "Cerebral hemodynamics during discrimination of prosodic and semantic emotion in speech studied by transcranial Doppler ultrasonography," *Neuropsychology* vol. 17, no. 1, January 2003.
[17] Gerd Gigirenzer, *Gut Feelings: The Intelligence of the Unconscious* (New York: Viking, 2007).
[18] Ito, "Digital Youth Project Findings."
[19] Mark Twain, *The Adventures of Tom Sawyer* (1876).
[20] "Is Multitasking More Efficient?" American Psychological Association, http://www.apa.org/releases/multitasking.html.
[21] Richard Mayer, Julie E. Heiser and Steve Lonn, "Cognitive constraints on multimedia learning: When presenting more material results in less understanding," *Journal of Educational Psychology* 93, no. 1 (March 1, 2001): 187–198.
[22] Walter Kirn, "The Autumn of the Multitaskers," *The Atlantic*, November 2007.
[23] John Hechinger, "U.S. Students Make Gains in Math Scores," *Wall Street Journal*, December 10, 2008.
[24] Bradshaw and Nichols, "Reading at Risk."
[25] Mark Bauerlein, *The Dumbest Generation: How the Digital Age Stupefies Young Americans and Jeopardizes Our Future* (New York: Tarcher, 2008).
[26] Nielsen Norman Group, "Usability of Websites for Teenagers," www.useit.com/alertbox/teenager.html., (January 31, 2005).

Chapter 4: Twittering Our Creativity Away

[1] Robert J. Sternberg, Ph.D., "President's Column: Creativity is a decision." *Monitor on Psychology* 34, no. 10 (November 2003).

[2] John S. Dacey and Kathleen H. Lennon, *Understanding Creativity: The Interplay of Biological, Psychological, and Social Factors* (Hoboken, NJ: Jossey-Bass, 1998).

[3] Shunryu Suzuki, *Zen Mind, Beginner's Mind* (Boston: Shambhala Publications, 2006).

[4] Rolex Mentor and Protégé Arts Initiative, Rolexmentorprotege.com (accessed on August 29, 2009).

[5] Michael Ray and Rochelle Myers, *Creativity in Business* (New York: Main Street Books, 1988).

[6] S. Dingfelder, "Creativity killers," *Monitor on Psychology* 34, no.10 (November 2003).

[7] Malcolm Gladwell, *Blink: The Power of Thinking Without Thinking* (New York: Little, Brown and Company, 2005).

[8] David Halberstam, *The Coldest Winter: America and the Korean War* (New York: Hyperion, 2007).

[9] A.J. Toynbee, *On the Future of Art* (New York: Viking Press, 1971).

[10] Palfrey and Gasser, *Born Digital: Understanding the First Generation of Digital Natives.*

[11] Meredith Etherington-Smith, *The Persistence of Memory: A Biography of Dali* (Cambridge, MA: Da Capo Press, 1995).

[12] Warren Brown, *Rockne* (Chicago: Reilly and Lee, 1931).

Chapter 5: Lost in the Game World

[1] Ito, "Digital Youth Project Findings."

[2] Evan Ramstad, "Coming to Tiny Screens All Over the Place," *The Wall Street Journal,* December 8, 2008.

[3] James Paul Gee, *What Video Games Have to Teach Us About Learning and Literacy* (Hampshire, United Kingdom: Palgrave Macmillan, 2007).

[4] David Brooks, "What Life Asks of Us," *The New York Times,* January 26, 2009.

[5] Gilbert Keith Chesterton, *Orthodoxy* (Sioux Falls: NuVision Publications, 2007).

[6] Ito, "Digital Youth Project Findings."

[7] Gee, *What Video Games Have to Teach Us About Learning and Literacy.*

[8] Ibid.

[9] Tom Drummond, "A Brief Summary of the Best Practices in Teaching," North Seattle Community College, http://webshares.northseattle.edu/eceprogram/bestprac.htm

[10] Ibid.

[11] Edward Shih-Tse Wang, et al., *Adolescence* 43 (March 22, 2008).

[12] Maria De Jong and Adriana G. Bus, "Quality of Book Reading Matters for Emergent Readers: An Experiment with the Same Book in Regular or Electronic Format," *Journal of Educational Psychology* 94, no.1 (March 2002): 145-155.

[13] Lawrence McCluskey, "Gresham's Law, Technology and Education," *Phi Delta Kappan* 75, no. 7 (March 1994): 550-552, http://www.questia.com/.

[14] Mack R. Hicks, Herbert Goldstein and Patricia Pearson, *Parent, Child and Community: A Guide for the Middle Class Urban Family* (Lanham, MD: Rowman & Littlefield Publishers, 1979).

[15] Gee, *What Video Games Have to Teach Us About Learning and Literacy*.

[16] Patricia Greenfield and Sandra Calvert, "Electronic media and human development: The legacy of Rodney R. Cocking," *Journal of Applied Developmental Psychology* 25 (November–December 2004): 627-631.

[17] Ito, "Digital Youth Project Findings."

[18] Ben Charny and Yukari Iwatani Kane, "Creator of Sims Games to Leave Electronic Arts," *The Wall Street Journal*,April 9, 2009.

[19] Mike Snider, "A Triple Play Awaits Fans of Sci-fi Gaming," *USA Today*, November 3, 2008.

[20] Small and Vorgan, *iBrain*.

[21] Ibid.

[22] Richard Restak, *Mozart's Brain and the Fighter Pilot: Unleashing Your Brain's Potential* (New York: Three Rivers Press, 2002).

[23] Torkel Klingberg, *Overflowing Brain: Information Overload and the Limits of Working Memory* (New York: Oxford University Press, 2008).

[24] Sadie F. Dingfelder, "Pitch Perfect," *Monitor on Psychology* 36, no.2 (February 2005): 32.

[25] Richard De Lisi and Jennifer L. Walford, "Improving Children's Mental Rotation Accuracy with Computer Game Playing," *Journal of Genetic Psychology* 163 (September 2002): 272-282.

[26] Thomas N. Robinson, et. al., "Effects of Reducing Children's Television and Video Game Use on Aggressive Behavior," *Archives of Pediatric and Adolescent Medicine* 155, no. 1 (January 2001): 17-23.

[27] Small and Vorgan, *iBrain*.

[28] David L. Strayer and William A. Johnston, "Driven to Distraction: Dual-Task Studies of Simulated Driving and Conversing on a Cellular Telephone," *Psychological Science* 12 (2001): 462-466.

Chapter 6: The Vital Regimen

[1] Palfrey and Gasser, *Born Digital.*

[2] Oscar Ybarra, "Mental Exercising Through Simple Socializing," *Personality and Social Psychology Bulletin* 34, no. 2 (February 2008).

[3] Chang-Hoan Cho and Honsik John Cheon, "Children's exposure to negative Internet content: effects of family context," *Journal of Broadcasting & Electronic Media* 42 (December 2005).

[4] Wang, et al., "The Relationship between leisure satisfaction."

[5] Jessica E. Vascellaro, "OMG, We're Not BFF's Anymore: Getting 'Unfriended' Online Stings," *The Wall Street Journal*, December 24, 2008.

[6] Brad Stone, "Is Facebook Growing Up Too Fast?" *The New York Times*, March 29, 2009.

[7] Elisabeth Parker, "Sexting—sending risque phone photos—landing teens in trouble,"*Tampa Tribune*, February 22, 2009.

[8] Dionne Searcey, "A Lawyer, Some Teens and a Fight Over 'Sexting'," *The Wall Street Journal*, April 21, 2009.

[9] Christine Rosen, "Machines That Won't Shut Up," *The Wall Street Journal*, April 12, 2009.

[10] Donna Winchester, "Cyberbullying on the rise," *St. Petersburg Times*, February 6, 2009.

[11] Alexandra Zayas, "Economy brings an end JuicyCampus Web site," *St. Petersburg Times*, February 6, 2009.

[12] Jessica E Vascellaro, "OMG, We're Not BFFs Anymore? Getting 'Unfriended' Online Stings," *The Wall Street Journal*, December 24, 2008.

[13] Lourdes Long, "Electronics as an extension of our physical selves," *Notre Dame Magazine*, Winter 2008-2009.

[14] Pavica Sheldon, "The relationship between unwillingness to communicate and students' Facebook use," *Journal of Media Psychology* 20, no. 2 (2008): 67-75.

[15] Susannah R. Stern, "Expression of Identity Online: Prominent features and gender differences in adolescents' World Wide Web home pages," *Journal of Broadcasting and Electronic Media* 48 (June 1, 2004).

[16] Hicks, Goldstein and Pearson, *Parent, Child and Community.*

[17] Ibid.

[18] Justin Kruger, Nicholas Epley, Jason Parker and Zhi-Wen Ng, "Egocentrism Over E-Mail: Can We Communicate as Well as We Think?" *Journal of Personality and Social Psychology* 89, no.6 (2005): 925-936.

[19] Frank A Drews, Monisha Pasupathi and David Strayer, "Passenger and Cell Phone Conversations in Simulated Driving," *Journal of Experimental Psychology* 14, no. 4 (2008): 392-400.

[20] Kameel Stanley, "A High-Tech Roadside Distraction," *St. Petersburg Times*, April 21, 2009.

[21] Victoria Rideout, "Parents, Media and Public Policy: A Kaiser Family Foundation Survey," The Henry J. Kaiser Foundation, Fall 2004, http://www.kff.org/entmedia/7156.cfm.

[22] Elizabeth Vandewater, et al. "Digital Childhood: Electronic Media and Technology Use Among Infants, Toddlers and Preschoolers," *Pediatrics* 119, no. 5 (May 2007): 1006-1015.

[23] D.R. Anderson and T. A. Pempek, "Television and Very Young Children," *American Behavioral Science* 48, no. 5 (2005): 505-522.

[24] Dimitri A. Christakis, et. al., "Early Television Exposure and Subsequent Attentional Problems in Children," *Pediatrics* 113, no. 4 (April 2004): 708-713.

[25] Judith Owens, et al., "Television-viewing Habits and Sleep Disturbance in School Children," *Pediatrics* 104, no. 3 (September 1999): 27.

[26] Thomas N. Robinson, et al., "Effects of Reducing Children's Television and Video Game Use on Aggressive Behavior."

[27] Werner Hopf, Gunter Huber and Rudolf Weib, "Media violence and youth violence: A 2-year logitudinal study," *Journal of Media Psychology: Theories, Methods and Applications* 20, no. 3 (2008): 79-96.

[28] "Violent Video Games Can Increase Aggression," American Psychological Association, http://www.apa.org/releases/videogames.html.

[29] Dr. Louise Hardy, "My Overweight Child: Tips, Strategies and Guidance for Parents of Overweight Kids." MyOverweightChild.com (accessed July 3, 2009).

[30] John M. Grohol, "Is a Digital Lifestyle a Deadly One?" PsychCentral, December 2, 2008, http://psychcentral.com/blog/archives/2008/12/02/is-a-digital-lifestyle-a-deadly-one/.

[31] Robert Bazell, "Too Much TV Could Cause Asthma," *NBC News*, NBC, August 24, 2009.

[32] George Will, "Let the Kids Go Outside and Play," *St Petersburg Times*, January 11, 2009.

[33] "Studies find clogged arteries in obese kids," *St. Petersburg Times*, November 12, 2008.

[34] Robert Larose, "Understanding Internet Usage," *Social Science Computer Review* 19, no. 4 (2001): 395-413, http://online.sagepub.com

Chapter 7: Education, Relaxation and Games

[1] Federation of American Scientists, "Summit on Educational Games: Harnessing the Power of Video Games for learning," Federation of American Scientists, October 2005, http://www.fas.org/gamesummit/.

[2]. Ibid.

[3] Sue Shellenbarger, "Playing Nice: Teachers Learn to Help Kids Behave in School," *The Wall Street Journal*, April 8, 2009.

[4]A.R. Luria, *The Man with a Shattered World* (Cambridge, MA: Harvard University Press, 1987).

[5] Patricia Lambert, "Imagination, Creativity and Frontal Lobe Function," personal interview with author, Pinellas Park, FL: Center Academy School, February 18, 2009.

[6] Peter Walker, "Education Matters," *RSA Journal* (Winter 2008): 12.

[7] Ito, "Digital Youth Project Findings."

[8] Rita Farlow, "Ex-Tarpon Springs High resource officer faces new accusation of misconduct at school," *St. Petersburg Times*, March 12, 2009.

[9] Michelle Galley, "Research: Boys to Men," *Education Week*, January 23, 2002.

[10] John A. O'Brien, Ph.D., *Cathedral Basic Readers—A Revision of Book One by William H. Elson and William Gray* (Chicago: Scott, Foresman and Co., 1931).

[11] Hicks, Goldstein and Pearson, *Parent, Child and Community.*

[12] Sue Shellenbarger, "How Not to Get Into College: Submit a Robotic Application," *The Wall Street Journal*, December 23, 2008.

[13] Cecil Cheek, personal interview with author, St. Petersburg College, March 11, 2009.

Chapter 8: Robotic or Romantic?

[1] Krueger, et.al. "Egocentrism Over E-mail: Can We Communicate As Well As We Think?"

[2]Lyric Wallwork Winik, "Intelligence Report: Safety," *Parade Magazine*, June 15, 2008.

[3] Patricia Evans, *Controlling People: How to Recognize, Understand, and Deal with People Who Try to Control You* (Cincinnati, OH: Adams Media, 2003).

Chapter 9: Contamination

[1] George Anders and Alan Murray, "Boardroom Duel: Behind H-P Chairman's Fall, Clash With a Powerful Director," *The Wall Street Journal*, October 9, 2006.

[2] Ibid.

[3] Tom Bower, "Hijacking a Dream," *The Daily Mail*, September 22, 2002.

[4] Alicia C. Shepherd, "The scoop on Woodward and Bernstein," *Los Angeles Times*, November 26, 2006.

[5] Peggy Noonan, "Rectitude Chic," *The Wall Street Journal*, December 12, 2008.

[6] Alan Murray, *Revolt in the Boardroom: The New Rules of Power in Corporate America* (New York: Harper Collins, 2007).

[7] Richard Willing, "Business with a Greater Purpose," *Notre Dame Business* (Winter 2005).

[8] Garrison Keillor, "News from Lake Wobegon," *Minnesota Public Radio*, September 29, 2007.

[9] Edmund S. Phelps, "Dynamic Capitalism," *The Wall Street Journal*, October 10, 2006.

Chapter 10: Disconnection in the Workplace

[1] Lesley Stahl,"How Technology May Soon Read Your Mind," *60 Minutes*, CBS, January 4, 2009.

[2] Clive Thompson, "Meet the Life Hackers," *The New York Times*, October 16, 2005.

[3] Brian Landman, "This season could be the last for Florida State Seminoles defensive coordinator Mickey Andrews," *St. Petersburg Times*, August, 10, 2009.

[4] Paul Paulus, *Group Creativity: Innovation through Collaboration* (New York: Oxford University Press, 2003).

[5] Jared Sandberg, "Colleagues You Wish You Didn't Have Seem the Most Secure," *The Wall Street Journal*, October 10, 2006.

[6] Phared Dvorak, "Corporate Meetings Go Through a Makeover," *The Wall Street Journal*, March 6, 2006.

[7] Robert Greenberg, *How to Listen to and Understand Great Music*, 3rd Edition (2004).

[8] The Teaching Company Conrad DeAenille, "Digital Archivists, Now in Demand," *The New York Times*, February 8, 2009.

[9] Ron Alsop, "The 'Trophy Kids' Go to Work," *The Wall Street Journal*, October 21, 2008.

Chapter 11: Gender Catharsis

[1] Daniel Ruth, "Yes Dear, You're Absolutely Right," *The Tampa Tribune*, December 3, 2000.

[2] Tom Wolfe, *A Man in Full* (New York: Farrar, Straus and Giroux, 1998).

[3] John Gray, *Men Are from Mars, Women Are from Venus* (New York: Harper-Collins, 1992).

[4] S. Pinker, *The Blank Slate: The Modern Denial of Human Nature* (New York: Viking, 2002).

[5] Elizabeth S. Spelke and D. Ariel Grace, "Abilities, Motives and Personal Styles," *American Psychologist* 61, no. 7 (October 2006).

[6] Eric Deggans, "One last round for the 'City' girls," *The St. Petersburg Times*, June 22, 2003.

[7] Dennis Prager, *CD #9: The Differences Between Men and Women*, (The Prager Perspective, LLC.)

[8] "I'm sorry for selling my story, says Iran hostage Mr. Bean," *London Evening Standard*, April 11, 2007.

[9] Turhan Canli, et al., "Sex differences in the neural basis of emotional memories," Proceedings of *The National Academy of Sciences* 99, no. 16 (2002).

[10] Louann Brizendine, *The Female Brain* (New York: Broadway Books, 2006).

[11] Richard Wiseman, "Reading Faces," Hertfordshire, Edinburgh International Science Festival, (Edinburgh, United Kingdom, 2005).

[12] Matthias Mehl, et. al., "Are Women Really More Talkative Than Men?," *Science* (July 2007).

[13] Tori DeAngelis, "Web pornography's effect on children," *Pediatrics* 119, no. 2 (February 2007): 247-257.

[14] Ito, "Digital Youth Project Findings."

[15] S.R. Stern, "Gender Differences in Style and Substance of Adolescent Personal Home Pages," (paper presented at the annual meeting of International Communication Association, San Diego, CA, May 27, 2003).

[16] D. Fallows, "How Women and Men Use the Internet" Pew Internet & American Life Project, December 2005, http://www.pewinter-net.org.

[17] Stern, "Expression of Identity Online."

[18] BA Shaywitz et al., "Sex differences in the functional organization of the brain," *Nature* (February 16, 195).

[19] Michael Price, "Lateral of the sexes," *Monitor on Psychology* 40, no. 1 (January 2009).
[20] Ann Moir and Jessel Davis, *Brain Sex: The Real Difference Between Men and Women* (New York: Dell Publishing, 1991).
[21] Ibid.
[22] Mark C. Gridley, "Cognitive Styles Partly Explain Gender Desparity in Engineering," *American Psychologist* 61, no 7 (October 2006).
[23] Spelke and Grace, "Abilities, Motives and Personal Styles."

Chapter 12: Physical Therapy

[1] Garry Trudeau, "Doonesbury," *St. Petersburg Times*, June 7, 2008.
[2] Eric Deggans, "Signal flair," *St. Petersburg Times*, May 25, 2007.
[3] Bill Marsh, "The Voice Was Lying. The Face May Have Told the Truth," *The New York Times*, February 15, 2009.
[4] Paul Ekman, *Emotions Revealed: Recognizing Faces and Feelings to Improve Communication and Emotional Life*, 2nd Ed. (New York: Henry Holt and Company, 2007).
[5] Stephen Young, "Hidden Meanings:In thousands of ways each day, many of us are unwittingly offending the people around us," *Town and Country Magazine* (August 2006).
[6] Mark Topkin, "Ideal Start for Maddon," *St. Petersburg Times*, March 3, 2006.
[7] William R. Levesque, "Lost a car, found a mess," *St. Petersburg Times*, December 23, 2007.
[8] Budd Davisson, "Measuring Up," *Flight Training Magazine*, June 2006.

Chapter 13: Interpersonal Politics

[1] Roger Simon, "So let the debates begin already. For Bush and Gore, it's put up or shut up time." *U.S. News and World Report*, October 2, 2000.
[2] Ibid.
[3] Elisabeth Bumiller, "Keepers of Bush Image Lift Stagecraft to New Heights," *The New York Times*, May 16, 2003.
[4] Roy P. Basler, ed. *Abraham Lincoln, His Speeches and Writings* (Cleveland and New York: The World Publishing Company, 1946).
[5] "Misunderestimated," *The Wall Street Journal*, June 28, 2006.
[6] Mark Shields, "Bush's Gaffes Shouldn't Be Cause for Glee," *The St. Petersburg Times*, September 17, 2000.
[7] Suzanne Sataline "Gov. Paterson Ends Senate-Seat Saga," *The Wall Street Journal*, January 23, 2009.
[8] Jan Morris, "!!!," *The Wall Street Journal*, June 28, 2006.

[9] Ann Blackman, "Campaign 2000: Take Note of Bob Graham," *Time*, July 17, 2000.

[10] Keith Epstein, "Trivial Pursuits: Is the note-taking obession Bob Graham shares with Thomas Jefferson a political liability?" *The Tampa Tribune*, July 13, 2003.

[11] Ibid.

[12] Drew Westen, *The Political Brain: The Role of Emotion in Deciding the Fate of the Nation* (New York: Public Affairs, 2007).

[13] Joe Queenan, "Hillary Is Too Boring to Be President," *The Wall Street Journal*, May 2008.

[14] "The Ronald Reagan Presidential Foundation and Literary Exhibit," (personal visit by author). Simi Valley, CA, November 12, 2006.

[15] Editorial, "My Heart and My Intentions," *St. Petersburg Times*, June 7, 2004.

[16] John Churchill, "In Wolfe's Clothing," *Phi Beta Kappa*, Summer 2006.

[17] Ibid.

[18] Nancy Kress, "Once More, From the Top," *Writer's Digest*, November 2005.

[19] Michael Gazzaniga, *Human: The Science Behind What Makes Us Unique* (New York: Harper Collins, 2008).

Chapter 14: Machines vs. Humanity

[1] Robert Ornstein, *The Right Mind: Making Sense of the Hemispheres* (New York: Harcourt Brace, 1997).

[2] Joseph Coates, "WorldView 2002! Futures Unlimited," (lecture presented at World Future Society's 22nd Annual Meeting, Philadelphia, PA, 2002).

[3] Erwin Chargaff, *Heraclitean Fire: Sketches from a Life Before Nature* (New York: Rockefeller University Press, 1978).

[4] William Hazlitt, trans., *The Table Talk or Familiar Discourse of Martin Luther* (Philadelphia: Lutherian Publication Society, 1997).

[5] Eric Deggans, "Cynical/Idealist," *The St Petersburg Times*, March 13, 2005.

[6] John Barry, "A scientist's brain scans to produce a photo album of the soul," *St. Petersburg Times*, June 13, 2008.

[7] Jorge Mall and Gordon Crofton, National Institute of Health, 2006.

[8] Edward O. Wilson, *Consilience: The Unity of Knowledge* (New York: Vintage, 1999).

[9] Wendell Berry, *Life Is a Miracle: An Essay Against Modern Superstition* (Berkley, CA: Counterpoint, 2001).

[10] Jeffrey Hammond, "Lost Souls," *Notre Dame Magazine* (Spring 2005).

[11] Chet Raymo, "The Eye of the Beholder," *Notre Dame Magazine* (Spring 2002).

[12] Eric Bern, International Transactional Analysis Association, www.itaa-net.org (accessed July 27, 2009).

[13] Calvin Hall and Gardner Lindzey, *Theories of Personality* (New York: John Wiley and Sons, 1957).

[14] Berry, *Life Is a Miracle.*

[15] *Man on Wire,* DVD, directed by James Marsh, (2008; Los Angeles, CA: Magnolia Home Entertainment, 2008).

[16] Lee Gomes, "The Singular Question Of Human vs. Machine Has a Spiritual Side," *The Wall Street Journal,* September 19, 2007.

Chapter 15: Future Mechanization Scenarios

[1] Charles R. Morris, "Genealogies of Morals," review of American Babylon by Richard John Neuhaus, *The New York Times Book Review*, April 10, 2009.

Chapter 16: From Impersonal Contact to Reconnection

[1] Christopher Lawton and Yukari Iwatani Kane, "Game Makers Push 'Family' Fare," *The Wall Street Journal,* October 29, 2008.

[2] Ito, "Digital Youth Project Findings."

[3] Drummond, "A Brief Summary of the Best Practices in Teaching."

[4] Creative Cities International, "Mission," Creative Cities International, LLC, http://www.creativecities.com/ mission.html (accessed July 17, 2009).

[5] Daniel H. Pink, *A Whole New Mind: Why Right-Brainers Will Rule the Future* (New York: Riverhead Books, 2006).

[6] Peggy Noonan, "Goodbye Bland Affluence," *The Wall Street Journal,* April 17, 2009.

Chapter 17: Suggestions for Inoculating the Family

[1] Sara Tucker, "Tech-Free in Moscow," *Conde Nast Traveler,* June 2009.

[2] Steven Johnson, "How the E-Book Will Change the Way We Read and Write," *The Wall Street Journal,* April 20, 2009.

[3] Tony Woodlief, "Ya Gotta Have (Real) Friends." *The Wall Street Journal,* June 12, 2009.

[4] Ibid.

[5] Elaine Birkinshaw, interview with author at Mirror Lake Branch Library, St. Petersburg, Florida, July 17, 2009.

Conclusion: Mechanization and the Hunter Spirit

[1] Vascellaro, "Facebook's About-Face on Data."

[2] Jacques Ellul, *The Technological Society*, trans. Robert K. Merton (New York: Vintage, 1967).

[3] Christine Rosen, "Machines That Won't Shut Up," *The Wall Street Journal*, April 10, 2009.

[4] Ellul, *The Technological Society.*

Appendix: Feed a Cold, Starve a Fever

[1] Grant M. Hill and Bud Turner. "A Checklist to Promote Physical Activity and Fitness." *The Journal of Physical Education, Recreation and Dance in K-12 Physical Education Programs* (November 2007).

[2] Ryan Knutson and Geoffrey A. Fowler. "Book Smarts? E-Texts Receive Mixed Reviews From Students." *Personal Journal, The Wall Street Journal* 78, no. 9, July 16, 2009.

[3] Patricia Lambert, personal interview with author, Pinellas Park, FL: Center Academy School, August, 2009.